The Open University

Mathematics/Science/Technology:

An Inter-Faculty Second Level Course

Mechanics and Applied Calculus

Unit 4 A FIELD OF FORCE AND SOME NUMERICAL METHODS OF SOLVING DIFFERENTIAL EQUATIONS

Unit 6 RIGID BODIES

Prepared by the Course Team

The Open University Press

Cover Photograph: Crown Copyright, Science Museum, London

The Open University Press, Walton Hall, Bletchley, Bucks.

First published 1972

Designed by the Media Development Group of the Open University.

Printed in Great Britain by
Martin Cadbury Printing Group

SBN 335 01172 1

This text forms part of the correspondence element of an Open University Second Level Course. The complete list of units in the course is given at the end of this text.

For general availability of supporting material referred to in this text, please write to the Director of Marketing, The Open University, Walton Hall, Bletchley, Buckinghamshire.

Further information on Open University courses may be obtained from the Admissions Office, The Open University, P.O. Box 48, Bletchley, Buckinghamshire.

1.1

Unit 4 A Field of Force and Some Numerical Methods of Solving Differential Equations

Contents

Bibliography

I. M. Khabaza, *Numerical Analysis* (Pergamon Press, 1965).

This book discusses basic numerical methods for solving first order differential equations, and comments on their usefulness.

Note

References to the Open University Mathematics Foundation Course Units (The Open University Press, 1971) take the form *Unit M100 3, Operations and Morphisms*.

Objectives

After working through this unit you should be able to:

(i) define the terms:

field of force,
gravitational force,
terminal speed,
uniform field;

(ii) distinguish between inertial and gravitational mass;

(iii) derive the parametric equations of a projectile in terms of parameter t for time, given appropriate initial conditions, for

(a) a uniform gravitational field of force with no air resistance;

(b) a uniform gravitational field of force with air resistance which is linearly dependent on the velocity (in still air);

(iv) use a numerical method to obtain an approximation to the path of the projectile when the air resistance does not depend linearly on the velocity.

Study Sequence

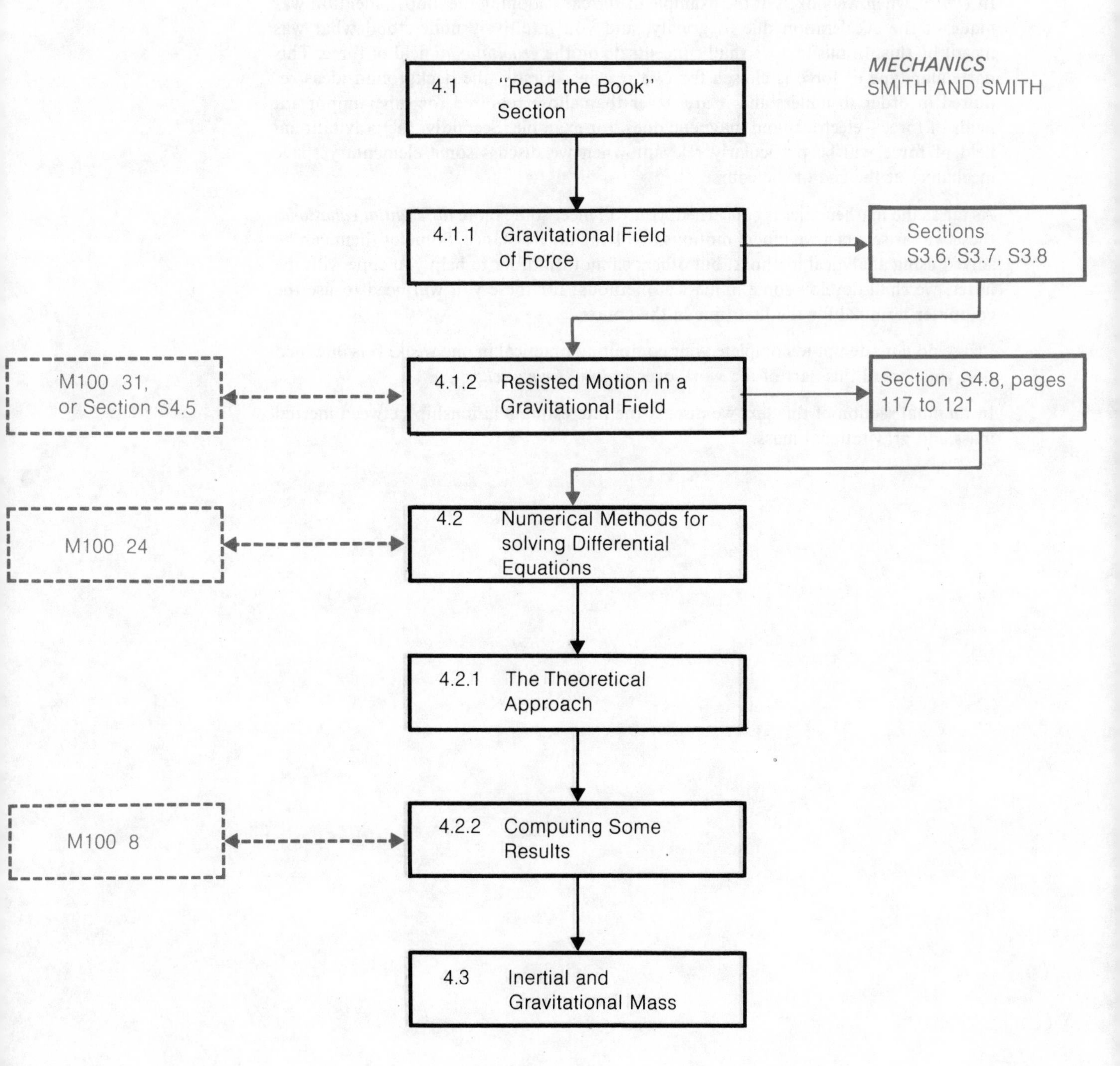

4.0 INTRODUCTION

In *Unit 3*, when we looked at the example of the car "looping the loop", mention was made of the acceleration due to gravity, and you intuitively understood what was meant by this. In this text we shall concentrate on the *gravitational* field of force. This particular type of force is chosen for two reasons. Firstly, the background ideas required in order to understand it are fewer than those required for other important fields of force—electrical and magnetic ones, for example. Secondly, the gravitational field of force will be particularly relevant when we discuss some elementary space mechanics at the end of the course.

As far as the mathematics is concerned, you will meet some more *differential equations*; these will arise when we model motions involving air resistance. Some of them can be solved using analytical methods, but others cannot. In order to help you cope with the latter, we shall develop some numerical methods; for these you will need to use the computer terminal for the first time in the course.

Please do *not* attempt to complete your computing practical in one week. It is intended that you spread this part of the work over *at least four weeks*.

In the final section of this text we discuss the interesting relationship between inertial mass and gravitational mass.

4.1 "READ THE BOOK" SECTION

4.1.1 The Gravitational Field of Force

The force due to gravity is curious in that it is not very strong. The gravitational attraction between a proton and an electron is 10^{29} times smaller than the electrical attraction between them. This is because gravitational force is related directly to the product of the masses involved; it is necessary to consider large masses before an appreciable gravitational effect is noticeable.

The earth is a body possessing a large mass; it exerts an appreciable force on an object close to it. This force, however, decreases when the distance between the object and the earth increases. Thus the gravitational force seems to depend both on the quantity of material that is causing it and also on the distance over which it is acting. The "action of a force over a distance" is a new concept in this course; it distinguishes gravitational force from the normal "push and pull" force of contact. Since it is difficult to observe any situation on or near the earth's surface which is not dominated by the great mass of the earth, it is remarkable that Newton noticed the mutual gravitational attraction between any two bodies. This aspect of gravitation is discussed in sections S3.6 and S3.7.

We shall examine the law of gravitational attraction, and then focus our attention upon a localized region of the force field that is sufficiently small to be considered uniform. We shall then examine the behaviour of a single particle (projectile) under the influence of such a field. Note that our initial mathematical model of the motion of the projectile is only a first approximation to the true motion, because we begin by ignoring air resistance. Later in this text we shall attempt to improve our mathematical model.

> Read sections S3.6, S3.7, S3.8 on pp. S59–69.
>
> There are several worked examples, particularly on pp. S65–69. We suggest you work through at least one of these, and more if you have time.

SAQ 1

Indicate the expression which completes the following sentence correctly.

The gravitational force on a point mass m_A at a point with position vector $\mathbf{r}_A$ due to a point mass m_0 at $\mathbf{r}_0$ is ...

A $\dfrac{\gamma m_A m_0(\mathbf{r}_A - \mathbf{r}_0)}{|\mathbf{r}_A - \mathbf{r}_0|}$

B $\dfrac{-\gamma m_A m_0(\mathbf{r}_A - \mathbf{r}_0)}{|\mathbf{r}_A - \mathbf{r}_0|^3}$

C $\dfrac{\gamma m_A m_0(\mathbf{r}_A - \mathbf{r}_0)}{|\mathbf{r}_A - \mathbf{r}_0|^2}$

D $\dfrac{\gamma m_A m_0(\mathbf{r}_A - \mathbf{r}_0)}{|\mathbf{r}_A - \mathbf{r}_0|^3}$.

(γ is the gravitational constant.)

(Solution is given on p. 27.)

SAQ 2

Exercise 16, p. S87.

(Solution is given on p. 27.)

4.1.2 Resisted Motion in a Gravitational Field

The next improvement in modelling the motion of particles in the vicinity of the earth's surface is to incorporate, to some extent, the force on the particle due to air resistance.

In attempting to formulate a mathematical model on the basis of experimental laws, we shall encounter equations which are not solvable by the usual analytic methods. This means that numerical techniques will be needed; we shall introduce some of these later in the text.

Read from the last paragraph on p. S117 to the bottom of p. S121.

If you wish, follow through the description of a numerical approach to the solution of a problem on pp. S122–124. (This last part can be left out if you are short of time.)

Again in reading the set book you will find two worked examples which you can treat as SAQ's. In the solutions given, the authors assume knowledge of second order differential equations which we have not yet discussed in this course. In fact, the equations involved can be written as first order differential equations, as in Example 18, p. S118, by writing $\frac{dz}{dt}$ as w. These differential equations can then be solved using techniques associated with first order equations. You will find such a solution to Example 18, p. S118, in the solution to SAQ 3 on p. 27 of this text. Alternatively, you could refer back to the treatment of second order linear differential equations with constant coefficients in *Unit M100 31*, *Differential Equations II*, or use section S4.5, pp. S101–109. We shall, however, look more fully at this type of equation later in the course.

SAQ 3

Example 18, p. S118. (See above.) (Solution is given on p. 27.)

(Solution is given on p. 27.)

SAQ 4

A particle cannot be moving at a speed greater than its terminal speed.
TRUE/FALSE.

(Solution is given on p. 28.)

SAQ 5

Raindrops hit the ground at about 7 ms^{-1} having fallen from a height of 4500 m. What would their speed be if there were no air resistance? Does this lead you to believe that the raindrops have effectively attained their terminal speed?

(Solution is given on p. 28.)

4.1.3 Summary

In section 4.1 we have introduced Newton's law of gravitation:

$$\mathbf{F} = -\frac{\gamma m_A m_B \mathbf{r}}{r^3},$$

which enables us to model the gravitational field of force generated by a body. Over a small area of the surface of a body as massive as the earth, we can approximate to the field of force by one that is constant in both magnitude and direction.

The first mathematical model we considered used this approximation; the differential equations arising from Newton's second law could be integrated to produce standard formulas for projectiles. In section 4.1.2 we improved the model by incorporating a term to represent air resistance; this term was taken to be proportional to velocity.

You should now be able to solve problems involving projectiles, including those with air resistance obeying this linear law, given appropriate initial conditions.

4.2 NUMERICAL METHODS FOR SOLVING FIRST ORDER DIFFERENTIAL EQUATIONS

4.2.0 Introduction

If, in the mathematical model of the motion of the projectile, the air resistance term is *not* proportional to velocity, then numerical methods may be needed to solve the differential equation. In this section we shall discuss a few of these methods. At the end of the section we have provided a planned set of exercises for you to do on the computer. This is not the first time you have met such numerical techniques. The Euler method was discussed in *Unit M100 24, Differential Equations I*; it is the basis of the technique used in Example 20, p. S122. In section 4.2.1 we remind you of the main points and extend the basic ideas.

4.2.1 The Theoretical Approach

Provided f is a sufficiently well-behaved function, the solution of a first order differential equation

$$\frac{dy}{dx} = f(x, y)$$

satisfying the initial condition

$$y = y_0 \quad \text{when} \quad x = x_0$$

is given by $y = y(x)$, where y is a continuous function. If it is possible to solve the differential equation simply by formula or analytic methods, then in practice we would do so. When using a numerical method, the best we can do is to evaluate y at a number of tabular points. Having found one tabular value, before we can find the next tabular value we must make some assumption about the behaviour of y in the interval between the two tabular points involved. The accuracy of the numerical technique depends on this assumption.

Generally, the smaller the interval between the tabular points, the more accurately the behaviour of the function in the interval can be predicted. Of course, there is a price to pay for increasing the number of tabular points—we must make a correspondingly increased number of calculations!

The given initial condition is that the graph of $y = y(x)$ must pass through the point (x_0, y_0). We shall write

$$x_r = x_0 + rh \qquad (r = 0, 1, 2, \ldots),$$

where h is the chosen constant spacing between the tabular points.

The corresponding tabular values are

$$y_r = y(x_r) = y(x_0 + rh) \qquad (r = 0, 1, 2, \ldots).$$

In the remainder of this section we shall list some of the basic numerical methods for solving first order differential equations, and comment on their usefulness. For a fuller description of these, turn, for example, to Khabaza, *Numerical Analysis* (see Bibliography).

A Taylor Series Method

In *Unit M100 14, Sequences and Limits II*, it was shown that, for a suitable function y whose successive derivatives at x_0 can be computed, the Taylor approximation of degree n can be used to evaluate the function at $x = x_0 + h$. The approximation is given by

$$y(x_0 + h) \simeq y(x_0) + hy'(x_0) + \frac{h^2}{2!}y''(x_0) + \frac{h^3}{3!}y'''(x_0) + \cdots + \frac{h^n}{n!}y^{(n)}(x_0).$$

But $y(x_0)$ is the given initial value and $y'(x_0) = f(x_0, y_0)$. So $y'(x_0)$ can be calculated by substituting values for x_0 and $y(x_0)$ in the given equation $y'(x) = f(x, y)$. The higher derivatives (at x_0) are found by successively differentiating the differential equation

$$y'(x) = f(x, y)$$

and substituting the values obtained in previous iterations, as we demonstrate in the following example.

Example

With a tabular interval $h = 0.1$, estimate the values of $y = y(x)$ at $x = 0.1, 0.2, \ldots, 0.8$, given that

$$y'(x) = \frac{dy}{dx} = x + y,$$

with initial values $x_0 = 0$, $y_0 = 1$.

This differential equation can be solved by the "integrating factor" method, but, despite our earlier remark, we shall use a numerical technique here for illustrative purposes.

Solution

We have

$$y'(x) = f(x, y) = x + y \quad \text{and} \quad h = 0.1,$$
$$y_0 = y(x_0) = y(0) = 1.$$

Now

$$y_1 = y(x_0 + h) = y(x_0) + hy'(x_0) + \frac{h^2}{2!}y''(x_0) + \frac{h^3}{3!}y'''(x_0) + \cdots$$

We obtain

$$y'(x_0) = x_0 + y_0 = 0 + 1 = 1;$$
$$y''(x) = 1 + y'(x),$$
$$y''(x_0) = 1 + y'(x_0) = 2;$$
$$y'''(x) = y''(x),$$
$$y'''(x_0) = y''(x_0) = 2;$$

and so on.

It follows that

$$y_1 = y(0.1) \simeq 1 + 0.1 \times 1 + 0.005 \times 2 + 0.000167 \times 2 = 1.1103$$

(The other terms in the Taylor series are so small that we can ignore them.) To estimate $y(0.2)$ we now have a choice of either

(i) setting $h = 0.2$, using the same initial condition as before (at $x = 0$, $y = 1$) and calculating $y(0.2)$,

or

(ii) using a new origin at $x = x_1$ with initial condition $y = 1.1103$

In choosing between the two approaches, we must balance the following factors. In (i) we have already evaluated $y'(x_0)$, $y''(x_0)$ etc., but we need to re-calculate $\frac{h^2}{2!}, \frac{h^3}{3!}$ etc.; furthermore, since h^r increases as h increases, we shall need more terms in the Taylor series to maintain the same accuracy. On the other hand, in (ii) we have calculated $\frac{h^2}{2!}, \frac{h^3}{3!}$ etc. but we need to calculate $y'(x_1)$, $y''(x_1)$ etc. The decision therefore rests to a considerable extent on the form of $f(x, y)$, that is, on whether or not it has a simple algebraic expression for the derivative. In general, (ii) provides the more economical approach. We shall use (ii) in this example. So we choose

$$x_0 = 0.1, \quad y_0 = 1.1103, \quad h = 0.1$$

Now

$$y'(x) = x + y, \quad \text{so} \quad y'(0.1) = 0.1 + 1.1103 = 1.2103$$
$$y''(x) = 1 + y'(x), \quad \text{so} \quad y''(0.1) = 1 + 1.2103 = 2.2103$$
$$y'''(x) = y''(x), \quad \text{so} \quad y'''(0.1) = 2.2103,$$

and so on. We obtain

$$y_2 = y(0.2) = 1.1103 + 0.1210 + 0.0111 + 0.0004 = 1.2428$$

For comparison, we note that the formula solution is

$$y = 2e^x - x - 1$$

and this gives $y_2 = 1.24281$

Similarly, we obtain

$$y_3 = 1.3997 \quad (1.39972)$$
$$y_4 = 1.5837 \quad (1.58365)$$
$$y_5 = 1.7974 \quad (1.79744)$$
$$y_6 = 2.0442 \quad (2.04424)$$
$$y_7 = 2.3275 \quad (2.32751)$$
$$y_8 = 2.6511 \quad (2.65108)$$

(The formula solution is given on the right for comparison.)

Let us look at the assumptions we have made about y in any particular interval. If we use only the first n terms of the Taylor series in our calculations, we are effectively assuming that all the remaining terms are zero. This means that the nth and all higher order derivatives of the function are assumed to be zero. This is true if the function is a polynomial of degree $(n - 1)$. In other words, we are assuming that y can be approximated (in a given interval of length h) by a polynomial of degree $n - 1$. This means that we are assuming that $\dfrac{h^n y^{(n)}}{n!}$ and higher terms are small enough not to matter.

B The Euler Method

The next method we investigate is one we met in *Unit M100 24*. The Euler method is a particular case of a Taylor series method, and involves taking only the first two terms of the series. Geometrically, we assume that in an interval h, y can be approximated by a straight line whose gradient is the same as that of the tangent to the graph of y at one end-point of the interval.

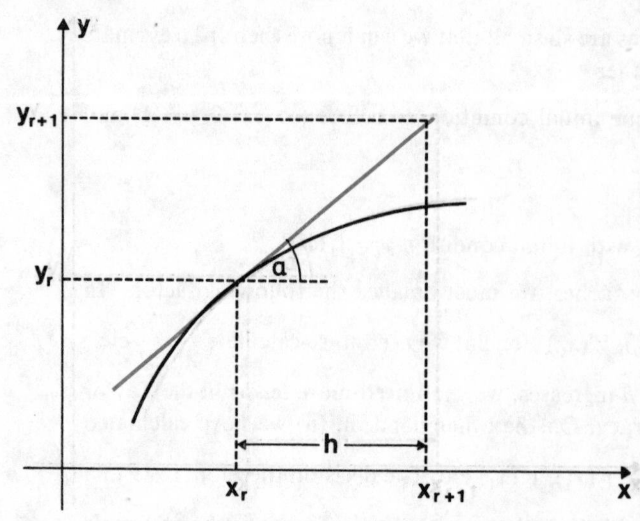

The gradient of this straight line is

$$\tan \alpha = \frac{y_{r+1} - y_r}{x_{r+1} - x_r} = \frac{y_{r+1} - y_r}{h}.$$

Thus

$$y_{r+1} = y_r + h \tan \alpha.$$

But

$$\tan \alpha = \frac{dy}{dx} \text{ at } (x_r, y_r) = f(x_r, y_r);$$

therefore

$$y_{r+1} = y_r + hf(x_r, y_r).$$

An application of this method to simultaneous first order differential equations is shown in Example 20, p. S122. (A more accurate method is subsequently used in this example.)

C Predictor-Corrector Methods

There are a number of Predictor-Corrector methods. The basis of these methods is that they proceed as their name suggests—they first obtain an estimate of y_{r+1}, and then this piece of information is used to make a better estimate. The methods can be regarded as two-step iterative methods and require two formulas. The first, known as the *predictor*, estimates y_{r+1} in terms of y_r. The other formula, known as the *corrector*, is used to make a better estimate of y_{r+1}, using both the given information (i.e. the value of y_r) *and* either the estimated value of y_{r+1}, or some other information obtained with the help of this estimated value.

For example, in the *modified Euler method* we use the Euler formula as a predictor to obtain the first approximation to y_{r+1}, denoted by $y_{r+1}^{(1)}$:

$$y_{r+1}^{(1)} = y_r + hf(x_r, y_r). \qquad \text{(i)}$$

This predictor formula gives only a rather crude estimate. The inaccuracy arises firstly because the linear approximation is used, and secondly because the slope of the approximating line is taken to be that of the curve at an end-point of the interval, $x = x_r$, rather than an average taken over the range of the interval.

Thus the predictor gives an approximate value of $y_{r+1}^{(1)}$. Hence we can evaluate (from the given differential equation) an approximate slope of the solution curve at $x = x_{r+1}$; it is $f(x_{r+1}, y_{r+1}^{(1)})$. Intuitively, a better estimate for y_{r+1} would be obtained if the approximating line had a slope which is the average slope of the two estimates at $x = x_r$ and at $x = x_{r+1}$, i.e. having a slope

$$\frac{f(x_r, y_r) + f(x_{r+1}, y_{r+1}^{(1)})}{2}.$$

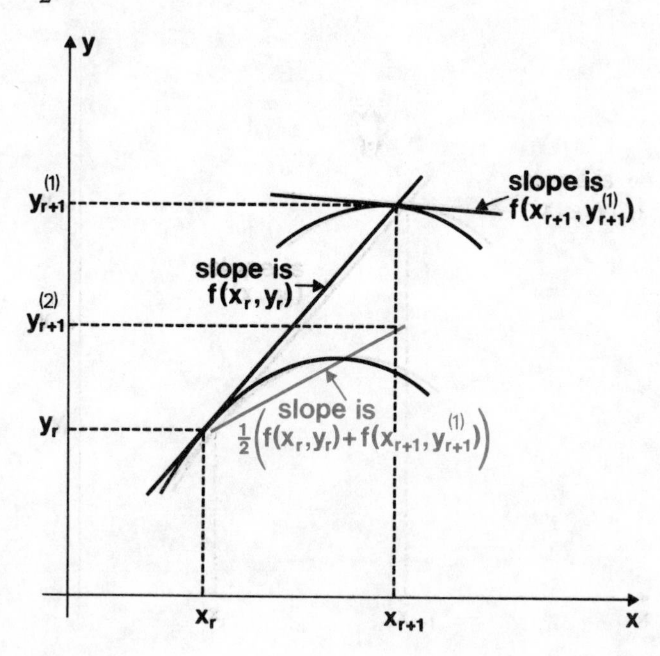

Using this straight line, we get a new estimate for y_{r+1}, namely

$$y_{r+1}^{(2)} = y_r + \frac{h}{2}\left(f(x_r, y_r) + f(x_{r+1}, y_{r+1}^{(1)})\right). \qquad \text{(ii)}$$

The $y_{r+1}^{(2)}$ on the left-hand side is the new corrected value; it will generally be different from (and more accurate than) the value obtained from equation (i) alone. Equation (ii) is known as the *corrector* formula.

Generally speaking, frequently used predictor-corrector methods, such as Milne's method and Adams-Bashforth's method, make use of more complicated formulas for the predictor and for the corrector. As far as this course is concerned, our objective is to give you the basic ideas behind the techniques, and not necessarily to discuss the most efficient numerical techniques.

D Other Methods

There are a number of other numerical methods used for solving differential equations. Among the best known is the *Runge-Kutta method*. One of the disadvantages of the Taylor series method is the necessity to compute higher order derivatives if we go beyond the Euler method. This is avoided in the Runge-Kutta method which instead uses several values of the given function $f(x, y)$. For the present we shall not develop this method; if you are interested, you will find it in most books on Numerical Analysis.

Summary

In this section we have discussed some basic numerical methods of solving first order differential equations. In the next section we ask you to practice some of these methods using a computer terminal.

4.2.2 Computing Some Results

This section contains three computer programs appropriate to three of the numerical methods we described in section 4.2.1. Our objective is to enable you to solve first order differential equations using your computer terminal and to encourage you to design suitable programs for yourself if you have the time. At the end of the section you are asked to compute the solution to Example 18, p. S118.

The programs are provided for (A) the Taylor series method, (B) the Euler method, (C) the modified Euler method. All programs are in BASIC.* The example used in each case is the worked example of the last section (page 13). You are advised to study the programs and compute the solution to each example given (or indeed, to any other similar examples). Choose an appropriate range for x, say $0 \leqslant x \leqslant 1$, and find the solutions to the equations by the three methods, choosing different values for h, say $h = 0.2$, 0.1, 0.05, 0.01 and 0.005.† Compare the results obtained by the different methods and try to account for discrepancies, if any!

In this section there are some changes in style. Firstly, we use Y_n instead of y_n for the tabular values of y, because we are now in a computing context. However, to avoid possible confusion, we retain the use of lower case letters for variables in quoted formulas (such as Taylor's approximation). Secondly, we have not put the solutions to the problems at the back of the text but have incorporated them with the programs.

A Taylor Series Method

Program Name: TYM282

The flow chart and the listing of this program are given in Appendix 1.

Program Description

The program solves a first order differential equation of the form

$$Y' = AX^B + CY^D + EX^FY^G + H,$$

using Taylor's series. A, B, C, D, E, F, G and H are constants.

The value of Y_{n+1} is calculated from the values $X = X_n$, $Y = Y_n$ using the fourth order Taylor series approximation:

$$y_{n+1} \simeq y_n + hy'_n + \frac{h^2}{2!}y''_n + \frac{h^3}{3!}y'''_n + \frac{h^4}{4!}y''''_n$$

(y'_n, y''_n, etc. are the derivatives of y at x_n).

Operating Instructions

The user must supply the values of A, B, C, D, E, F, G and H, and the initial conditions for X and Y. In addition, the step size and the upper limit of the range of X are required. For example, for the equation

$$Y' = X + Y,$$

we have

$$A = 1, B = 1, C = 1, D = 1, E = F = G = H = 0.$$

The step size and the upper limit of the range are chosen by the user.

* BASIC was developed at Dartmouth College, New Hampshire, U.S.A. by Professor J. G. Kemeny and Professor T. E. Kartz.

† Data to be inputted are printed inred.

Worked Example

Solution of

$$Y' = X + Y$$

```
GET—$TYM282

RUN

TYM282

PROGRAM TO EVALUATE THE SOLUTION OF A DIFFERENTIAL
EQUATION NUMERICALLY USING TAYLOR'S SERIES

EQUATION IS IN THE FORM:—

        Y' = A*X ↑ B + C*Y ↑ D + E*(X ↑ F)*(Y ↑ G) + H

PLEASE TYPE VALUES OF A, B, C, D, E, F, G, H?1, 1, 1, 1, 0, 0, 0, 0
PLEASE INPUT THE INITIAL CONDITIONS, I.E. VALUE OF
Y FOR STARTING VALUE OF X
TYPE VALUES FOR X AND Y?0,1
PLEASE TYPE IN UPPER BOUND FOR RANGE OF X?1
PLEASE TYPE STEP SIZE FOR X?0.05
```

XN	YN
0	1
.05	1.05254
.1	1.11034
.15	1.17367
.2	1.24281
.25	1.31805
.3	1.39972
.35	1.48813
.4	1.58365
.45	1.68662
.5	1.79744
.55	1.91651
.6	2.04424
.65	2.18108
.7	2.3275
.75	2.484
.8	2.65108
.85	2.82929
.9	3.0192
.95	3.22142
1.	3.43656

```
DONE
```

B Euler's Method

Program Name: EUM282

The flow chart and the listing of this program are given in Appendix 2.

Program Description

The program solves a first order differential equation of the form

$$Y' = AX^B + CY^D + EX^FY^G + H,$$

using the Euler approximation:

$$y_{n+1} = y_n + hy_n.$$

Operating Instructions

The user must supply the values of A, B, C, D, E, F, G and H, and the initial conditions for X and Y. In addition, the step size and the upper limit of the range of X are required. For example, for the equation

$$Y' = X + Y,$$

we have

$$A = 1, B = 1, C = 1, D = 1, E = F = G = H = 0.$$

The step size and the upper limit of the range are chosen by the user.

Worked Example

Solution of

$$Y' = X + Y$$

```
GET—$EUM282

RUN

EUM282

PROGRAM TO EVALUATE THE SOLUTION OF A DIFFERENTIAL
EQUATION NUMERICALLY USING EULER'S METHOD

EQUATION IS IN THE FORM:—

    Y' = A*X ↑ B + C*Y ↑ D + E*(X ↑ F)*(Y ↑ G) + H

PLEASE TYPE VALUES FOR A, B, C, D, E, F, G, H?1, 1, 1, 1, 0, 0, 0, 0
PLEASE INPUT THE INITIAL CONDITIONS, I.E. VALUE
OF Y FOR STARTING VALUE OF X.
PLEASE TYPE VALUES FOR X AND Y?0,1
PLEASE TYPE UPPER BOUND FOR RANGE OF X?1
PLEASE TYPE STEP SIZE FOR X?0.05

            XN              YN
            0               1
             .05            1.05
             .1             1.105
             .15            1.16525
             .2             1.23101
             .25            1.30256
             .3             1.38019
             .35            1.4642
             .4             1.55491
             .45            1.65266
             .5             1.75779
             .55            1.87068
             .6             1.99171
             .65            2.1213
             .7             2.25986
             .75            2.40786
             .8             2.56575
             .85            2.73404
             .9             2.91324
             .95            3.1039
            1.              3.3066

DONE
```

C Modified Euler Method

Program Name: EMM282
The flow chart and the listing of this program are given in Appendix 3.

Program Description

The program solves a first order differential equation of the form

$$Y' = AX^B + CY^D + EX^FY^G + H,$$

using a modified Euler method.

The value of Y_{n+1} is calculated from the values $X = X_n$, $Y = Y_n$ using the Euler approximation:

$$y_{n+1} = y_n + hy'_n \qquad \text{(predictor formula).}$$

A value of Y'_{n+1} is then found using the initial equation

$$Y' = AX^B + CY^D + EX^FY^G + H.$$

A corrected value of Y_{n+1} is given by

$$y_{n+1} = y_n + \frac{h}{2}(y'_n + y'_{n+1}) \qquad \text{(corrector formula).}$$

Operating Instructions

The user must supply the values of A, B, C, D, E, F, G and H, and the initial conditions for X and Y. In addition, the step size and the upper limit of the range of X are required. For example, for the equation

$$Y' = X + Y,$$

we have

$$A = 1, B = 1, C = 1, D = 1, E = F = G = H = 0.$$

The step size and the upper limit of the range are chosen by the user.

Worked Example

Solution of

$$Y' = X + Y$$

```
GET—$EMM282

RUN

EMM282

PROGRAM TO EVALUATE THE SOLUTION OF A DIFFERENTIAL
EQUATION NUMERICALLY USING A MODIFIED EULER METHOD.

EQUATION IS IN THE FORM:—

      Y' = A*X ↑ B + C*Y ↑ D + E*(X ↑ F)*(Y ↑ G) + H

PLEASE TYPE VALUES FOR A, B, C, D, E, F, G, H?1, 1, 1, 1, 0, 0, 0, 0
PLEASE INPUT THE INITIAL CONDITIONS, I.E. VALUE
OF Y FOR STARTING VALUE OF X
PLEASE TYPE VALUES FOR X AND Y?0,1
PLEASE TYPE UPPER BOUND FOR RANGE OF X?1
PLEASE TYPE STEP SIZE FOR X?0.05
```

XN	YN
0	1
.05	1.0525
.1	1.11025
.15	1.17353
.2	1.24261
.25	1.31779
.3	1.39939
.35	1.48774
.4	1.58317
.45	1.68606
.5	1.79678
.55	1.91574
.6	2.04336
.65	2.18008
.7	2.32637
.75	2.48273
.8	2.64965
.85	2.8277
.9	3.01743
.95	3.21945
1.	3.43438

DONE

SAQ 6

We want you to calculate an approximate numerical solution to Example 18, p. S118. The problem is:

> A man, falling with speed 50ms^{-1} at a height of 200m, opens his parachute. With resistance proportional to velocity, the terminal speed of the parachutist is 5ms^{-1}. Estimate his time of descent.

We suggest that you solve this problem in several stages, depending on how much time you have available.

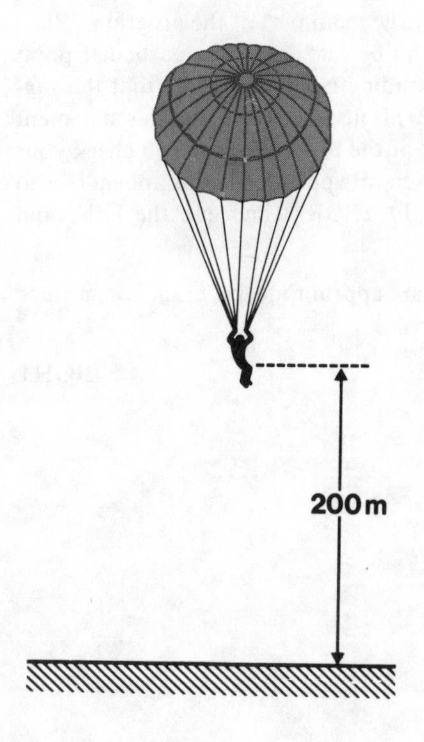

The equation of motion is

$$\frac{d^2z}{dt^2} = -g - k\frac{dz}{dt}$$

or, if w is the velocity,

$$\frac{dw}{dt} = -g - kw.$$

(a) Now try to solve this equation numerically for w using the modified Euler method program, EMM282 with step 0.2s, after determining k. Assume $g = 9.8\text{ms}^{-2}$.

(Solution is given on p. 28.)

(b) In part (a) you calculated the velocity w at intervals of 0.2s. Your print-out should show that the terminal speed is reached in about 10s.

Now write a short program, using the Euler method with step size 0.2s, to determine the height of the man after 10s. In this case the derivative you use at the beginning of each time interval will be the velocity at the beginning of that interval (which you calculated in part (a)). Thus the set of velocity values obtained through running your first program should now be used as initial data for your second program.

(Solution is given on pp. 29–31.)

(c) Now calculate by hand the approximate total time before the man reaches the ground. What assumption have you made?

(Solution is given on p. 31.)

(d) It is a rather long-winded way of solving this problem to use (a) and (b) consecutively. Try incorporating a subroutine using the modified Euler method to produce velocity and height, using the Euler method (not as a subroutine) at the same time.

First, try to modify EMM282 to produce a subroutine EUMSUB starting at line 5000. The idea of the subroutine is to allow a step by step solution of the differential equation rather than a complete solution tabulated over each interval in a given range. This means that the subroutine should assume that the parameters A, B, C, D, E, F, G and H (of the equation $Y' = AX^B + CY^D + EX^FY^G + H$) have been provided in the main body of the program, in addition to the step size (S) and the values of X and Y from which Y(X + S) is to be calculated. The subroutine should set the new value of Y in a specified location so that it may be used by the main part of the program. When a subroutine is used, it means that the velocity can be calculated at a particular point and used automatically to calculate the corresponding height. The fact that the suggested starting statement for the subroutine is 5000 is just to ensure that this statement is not confused with statements in the main part of the program. You can check your program with the program we have designed, given in Appendix 4; this appendix also includes two other subroutines, EULSUB and TAYSUB, which use the Euler and Taylor methods respectively.

Second, try to incorporate this subroutine in an appropriate program to produce tabulated values and headings:

TIME	VELOCITY	HEIGHT

(Solution is given on p. 31.)

4.3 INERTIAL AND GRAVITATIONAL MASS

In this final section we return to the subject of *mass*. The quantity *mass* can be viewed in two distinct ways.

First, it can be defined as the factor of proportionality between the force **F** and the acceleration **a** in Newton's second law of motion:

$$\mathbf{F} = m\mathbf{a}.$$

We can (theoretically) label the masses of all bodies by performing the "spring experiment" with them, as described in the Introduction to *Unit 3*, and on p. S51. Thus the ratio of the masses of two bodies can be specified by

$$\frac{m_1}{m_2} = \frac{a_1}{a_2},$$

where a_1 and a_2 are the accelerations produced in the respective masses by the same spring applying the same force to each mass. By this method, we can label all other masses by measurements made relative to one standard mass. This manner of defining mass is purely inertial in nature, because the applied force is independent of the body on which it acts. Such a mass is defined as the *inertial mass* of a body; we write it as m_I. Newton's second law is then written:

$$\mathbf{F} = m_I\mathbf{a}.$$

The second way of defining *mass* comes from another of Newton's laws, the law of gravitation, which states that the force of attraction between two bodies is proportional to some intrinsic property of each body. For the time being, let us call this property the *gravitational charge*. Experimentally, it is known that the force of gravity between two particles varies inversely as the square of the distance between them. In this respect the force of gravity is similar to electrostatic attraction. As you may know, electrostatic charges can be placed on particles, and these charges give rise to forces that are also inversely proportional to the square of the distance between the particles. However, there is an important difference between electrostatic and gravitational forces. Whereas the amount of electrostatic charge on a given body may be varied within limits at will (and thus the electrostatic force can be varied), the gravitational charge on the body cannot be changed without at the same time changing the inertial mass by a proportionate amount. Thus, for example, if the gravitational charge on the body is doubled, its inertial mass is also doubled, whereas a doubling of the electrostatic charge on the body does not necessarily lead to a similar change in inertial mass. (We leave you to contemplate the interesting consequences of being able to change the "gravitational charge" at will.)

The point about these two quantities, inertial mass and gravitational charge, is that they are always directly proportional to each other; for this reason it has become customary to use the term *gravitational mass* rather than gravitational charge; we write it as m_G. We shall now describe how this proportional relationship was investigated experimentally.

The magnitude of the force of gravity between the earth and a particle is given by

$$F_G = (\text{constant})\,\frac{m_G}{r^2},$$

where r is the distance between the particle and the centre of the earth, and m_G is the gravitational mass of the particle. If we are considering a localized area near the earth's surface, F_G is directed downwards with magnitude $F_G = km_G$, where k is some constant of proportionality.

What methods can be used to check whether the inertial mass m_I is proportional to the gravitational mass m_G?

One method is to use a pendulum made by suspending a bob of inertial mass m_I from a string of negligible mass and length l. If air friction is small enough to be ignored, the component of force in the direction of increasing ϕ is

$$F_\phi = -\,km_G \sin\phi.$$

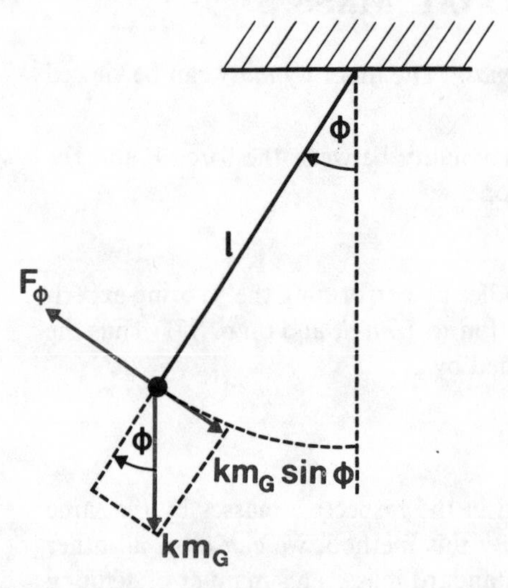

Note that the force is a restoring force, which attempts to restore the pendulum to a position of equilibrium. Since the length of the pendulum is fixed, the acceleration is

$$f_\phi = l\ddot{\phi}.$$

Here we are using the formula for the transverse component of acceleration, given in *Unit 2* (or on p. S19). The equation of motion is therefore

$$-km_G \sin\phi = m_I l\ddot{\phi}$$

or

$$\ddot{\phi} = -\frac{k}{l}\frac{m_G}{m_I}\sin\phi.$$

Now if m_G and m_I are strictly proportional, we have

$$\ddot{\phi} = (\text{constant})\sin\phi.$$

Thus, for fixed l, the motion of the pendulum, and hence the period of its oscillation, is the same for any bob–a remarkable fact!

In his famous book, *The Mathematical Principles of Natural Philosophy* (*Principia* for short), first published in 1687, Newton describes an experiment along these lines. Because *Principia* also contains his enunciation of the three laws, many important examples of such experiments are often overlooked. But their inclusion serves to emphasize that Newton was a careful experimental scientist as well as a mathematical genius.

Newton constructed two identical pendulum bobs from spherical wooden boxes. This ensured that the air resistance acting on them would be the same. He then filled the boxes with *different* materials such as lead, silver, glass, sand, wood or wheat. He showed, to an accuracy of more than one part in a thousand, that in all cases the periods of motion of the two boxes were the same.

An alternative method for investigating the relationship between m_I and m_G is that used more recently, in 1909, by Baron Eotvos in Budapest. It uses a rather clever static measurement that relies on the fact that, because of its rotation, the earth is *not* an inertial system. (We shall discuss rotating frames of reference in *Units 13* and *14*.)

To understand the principle, consider a simple pendulum hanging at rest near the earth's surface. Let θ be the latitude and T the tension in the string.

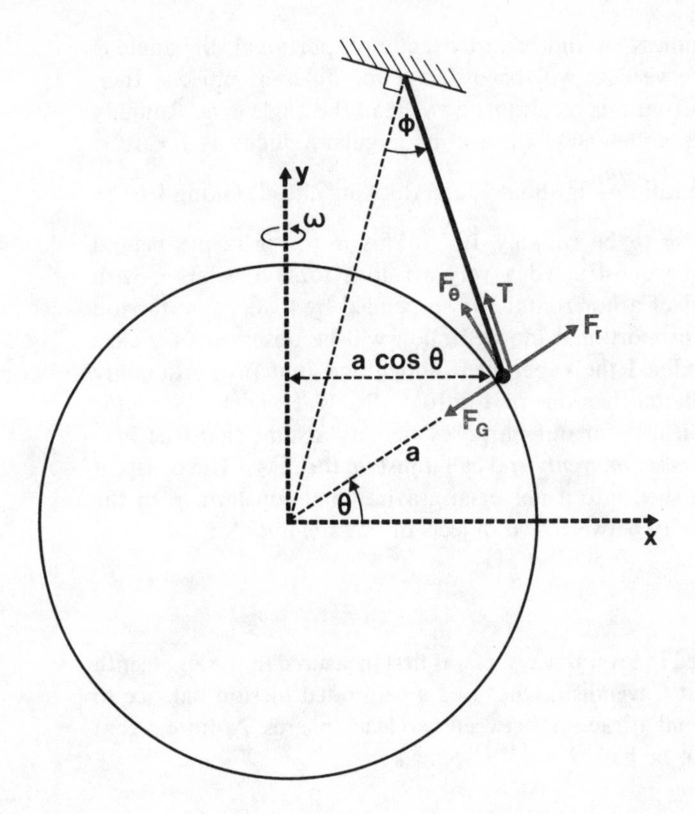

Because of the earth's rotation, the pendulum bob hangs slightly off the true vertical by an angle ϕ. (This will be explained in detail in *Units 13* and *14*.)

At the latitude θ on the earth's surface, the bob has radial component of acceleration $a\omega^2 \cos\theta$ in the direction of the negative x-axis, where the earth's rotation is characterized by the angular velocity ω, and a is the earth's radius. The accelerations and forces may be resolved in the directions indicated by the suffices θ and r.

The components of acceleration along the increasing θ-direction and the increasing r-direction are

$$f_\theta = a\omega^2 \cos\theta \sin\theta$$
$$f_r = -a\omega^2 \cos^2\theta.$$

The components of force along these directions are

$$F_\theta = T \sin\phi$$

and

$$F_r = T\cos\phi - F_G,$$

where $F_G\ (= km_G)$ is the force due to gravity. Now if these forces and accelerations are substituted in Newton's second law, $\mathbf{F} = m_I \mathbf{f}$, we obtain the equations

$$m_I a\omega^2 \cos\theta \sin\theta = T\sin\phi$$

and

$$-m_I a\omega^2 \cos^2\theta = T\cos\phi - km_G.$$

The angle ϕ is therefore given by

$$\tan\phi = \frac{m_I a\omega^2 \sin\theta\cos\theta}{km_G - m_I a\omega^2 \cos^2\theta}$$

$$= \frac{\dfrac{m_I}{m_G}\sin\theta\cos\theta\,\dfrac{a\omega^2}{k}}{1 - \dfrac{a\omega^2}{k}\dfrac{m_I}{m_G}\cos^2\theta}.$$

From this result we see that, unless m_I and m_G are exactly proportional, the angle ϕ of the deviation from the true vertical will be different for different objects. Incidentally, we can also estimate from this result just how small the angle ϕ is. Roughly speaking, the earth's radius is about 6000 km and its angular velocity is 7×10^{-5} radians s^{-1}. Assuming that the ratio $\frac{m_I}{m_G}$ is about 1, and (looking ahead) taking k to be about 9.8 ms^{-2}, we find $\tan \phi$ to be roughly 10^{-3}. This is the principle behind Eotvos's experiment. In fact he constructed a very sensitive torsion balance, with dissimilar objects at either end of a horizontal bar suspended by a very fine torsion fibre. If m_I and m_G are truly proportional, no deflection will be observed in such a balance. Dicke has recently refined the experiment and found that proportionality holds to an accuracy which is better than one part in 10^{10}. (See R. H. Dicke, *Scientific American 205*, (Dec. 1961).) Clearly, for our purposes we may assume that true proportionality holds. Indeed, let us set $m_I = m_G$ and call it just m, the mass. The constant of proportionality is now absorbed into a universal gravitational constant, γ, in the expression for the force of gravity between two objects of mass m and M:

$$\mathbf{F} = -\frac{\gamma M m}{r^3}\mathbf{r}.$$

The value of γ can be determined in several ways. It was first measured in the eighteenth century by the English scientist Cavendish, who used a calibrated torsion balance to measure directly the gravitational attraction between two lead spheres. A more recent (1942) value for γ was found to be 6.6732×10^{-11} $Nm^2\ kg^{-2}$.

SAQ 7

Measurements made with bathroom scales involve the use of a spring system. Indicate which of the following statements is/are correct.

A The scales record the inertial mass of a person standing on them.
B The scales record the gravitational mass of a person standing on them.
C The scales rely for their validity on the equivalence of inertial and gravitational mass.

(Solution is given on p. 31.)

4.4 SOLUTIONS TO SELF-ASSESSMENT QUESTIONS

Solution to SAQ 1

D

Solution to SAQ 2

35.3m, 5.1m.

The general case for a projectile with initial speed V at angle α to the horizontal is analysed on pp. S62–64, and you could substitute the particular quantities in the expressions derived there. If, however, you derived the expressions yourself, you should have covered the main steps summarized below.

The horizontal and vertical components of the initial velocity are $20\cos 30^\circ\ \mathrm{ms}^{-1}$ and $20\sin 30^\circ\ \mathrm{ms}^{-1}$ respectively.

The horizontal and vertical components of the velocity at subsequent time t are $20\cos 30^\circ\ \mathrm{ms}^{-1}$ and $(20\sin 30^\circ - gt)\ \mathrm{ms}^{-1}$ respectively.

The horizontal and vertical distances travelled at any time t (before the projectile hits the ground) are given by $x = 20t\cos 30^\circ$ and $z = 20t\sin 30^\circ - \frac{1}{2}gt^2$ respectively.

When the particle reaches the end of its range, the height is zero again. This occurs at time $t_0 = \dfrac{40\sin 30^\circ}{g}$.

Then x_0 (the horizontal range) = 35.3m.

The maximum height occurs when the vertical component of the velocity is zero, i.e. after time

$$t_1 = \frac{20\sin 30^\circ}{g}.$$

Then $z = 5.1$ m, by substitution.

Solution to SAQ 3 (Example 18, p. S118, alternative form)

Substituting $\dfrac{dw}{dt}$ in the second order differential equation on the 6th line of p. S119, we obtain

$$\frac{dw}{dt} = -g - kw, \qquad (k = 1.96\mathrm{s}^{-1}),$$

which is in "variables separable" form, so

$$\frac{dw}{g + kw} = -dt.$$

Integrating, we obtain

$$\frac{1}{k}\ln(g + kw) = -t + c,$$

so

$$g + kw = Ae^{-kt}, \qquad \text{(i)}$$

where A is a constant to be determined from the initial conditions. The initial speed is $50\ \mathrm{ms}^{-1}$ (downwards) and therefore $w = -50$ when $t = 0$. Using these conditions in (i), we obtain

$$A = 9.8 + 1.96(-50) = -88.2$$

Equation (i) therefore becomes

$$9.8 + 1.96w = -88.2e^{-1.96t}$$

If z is the height above the ground at time t, then the upward speed $w = \frac{dz}{dt}$, and therefore

$$9.8 + 1.96\frac{dz}{dt} = -88.2e^{-1.96t}$$

This can be written in "variables separable" form as

$$1.96\frac{dz}{dt} = -88.2e^{-1.96t} - 9.8$$

Integrating, we obtain

$$1.96z = 45e^{-1.96t} - 9.8t + B.$$

But $z = 200$ m when $t = 0$, and therefore

$$B = 392 - 45 = 347$$

It follows that

$$1.96z = 45e^{-1.96t} - 9.8t + 347 \qquad \text{(ii)}$$

The man reaches the ground when $z = 0$, and therefore the time taken is the solution of the equation

$$0 = 45e^{-1.96t} - 9.8t + 347$$

Notice that the first term decreases as t increases, and that when t is 3 seconds, for example, we have (using tables of exponential functions)

$$45e^{-5.88} = 45 \times 0.0028 \simeq 0.13,$$

which is small in comparison with the other terms.

After 3 seconds, the man has not yet reached the ground (substitution in (ii) will confirm this) and the first term will be even smaller when he eventually does so. We can therefore obtain an approximate solution by neglecting the first term, and solving the equation

$$-9.8t + 347 = 0.$$

This gives a first approximation of about 35 s for the time of descent.

If a better estimate is required for the time of descent, we could theoretically use the recurrence formula

$$t_{n+1} = \frac{347 + 45.7e^{-1.96t_n}}{9.8}.$$

But $e^{-1.96\times 35}$ is so small that it would not affect the accuracy to which we have been working in the solution.

Solution to SAQ 4

FALSE. The particle's initial speed could be greater than its terminal speed.

Solution to SAQ 5

The raindrop's final speed would be $\sqrt{2gh}$ (where h is the height fallen), which is $\simeq 300$ ms^{-1}. In fact, a raindrop effectively attains its terminal speed after a few tens of metres.

Solution to SAQ 6

(a) The terminal speed is determined by putting $\frac{dw}{dt} = 0$. Thus k is given by

$$0 = -9.8 - k(-5).$$

Hence $k = 1.96s^{-1}$. We therefore solve, using program EMM 282, the equation

$$\frac{dw}{dt} = -9.8 - 1.96w$$

(Y being w and X being t, with initial condition $w = -50$ when $t = 0$.) We assume, for convenience, that the man reaches his terminal velocity in less than 12 s. So we input an upper bound of 12 on the range of X (the time).

The program and solution are given below. Note that we have reproduced only a section of the printed pairs of values which should appear on your print-out.

```
GET—$EMM282

RUN

EMM282

PROGRAM TO EVALUATE THE SOLUTION OF A DIFFERENTIAL
EQUATION NUMERICALLY USING A MODIFIED EULER METHOD
EQUATION IS IN THE FORM:—
        Y' = A*X↑B + C*Y↑D + E*(X↑F)*(Y↑G) + H

PLEASE TYPE VALUES FOR A,B,C,D,E,F,G,H ?0,0,-1.96,1,0,0,0,-9.8

PLEASE INPUT THE INITIAL CONDITIONS, I.E. VALUE
OF Y FOR STARTING VALUE OF X

PLEASE TYPE VALUES FOR X AND Y?0,-50

PLEASE TYPE UPPER BOUND FOR RANGE OF X?12

PLEASE TYPE STEP SIZE FOR X?.2
```

XN	YN
0	-50
.2	-35.8174
.4	-26.1048
.6	-19.4532
.8	-14.898
1.	-11.7785
2.	-6.02106
3.	-5.1538
4.	-5.02317
5.	-5.00349
6.	-5.00053
7.99999	-5.00001
10.	-5.
12.	-5.

```
DONE
```

It now appears that the man reaches his terminal velocity after about 9 s.

(b) For a given interval, the difference in the heights at the beginning and end, Z_{n-1} and Z_n, is approximately equal to the velocity at the beginning of the interval, W_{n-1}, multiplied by the time 0.2. Thus the iterative formula for the Euler method becomes

$$Z_n = Z_{n-1} + 0.2W_{n-1}.$$

The program is listed below.

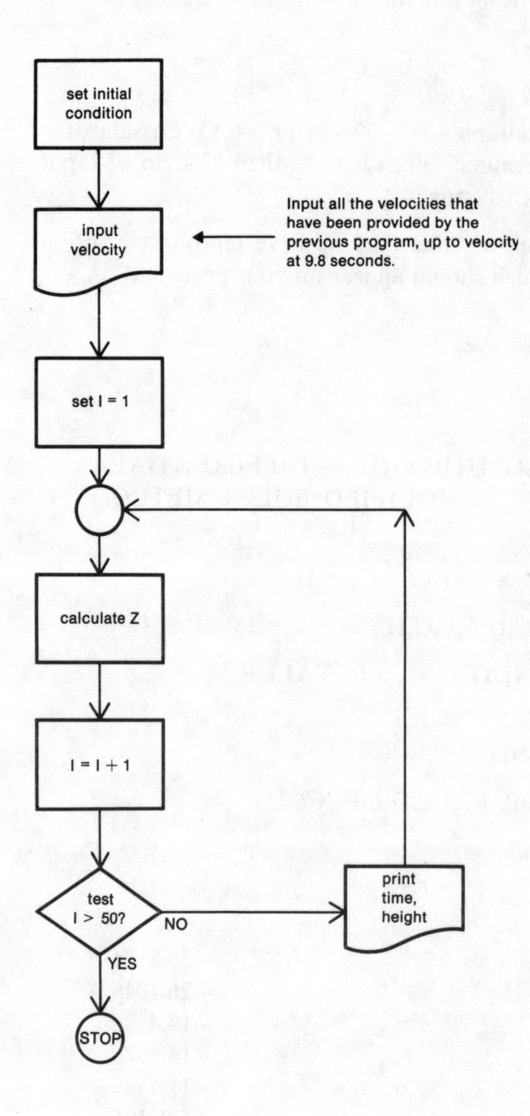


```
LIST
 5  DIM  A  (50)
10  LET  Z = 200
15  FOR  I = 1  TO  50
20  INPUT  A  (I)
25  NEXT  I
30  PRINT  "T", "ZN"
35  FOR  I = 1  TO  50
40  LET  Z = Z + .2*A  (I)
45  PRINT  I/5,  Z
50  NEXT  I
55  END
```

Selected values from the output are listed below.

```
RUN

?-50

?-35.8174

  T                                  ZN
 .2                                 190
 .4                                 182.837
```

1	170.745
5	146.446
10	121.444

DONE

Thus the approximate height after 10 s is 120 m.

(c) The assumption is that the man falls for the rest of the distance at the terminal speed, 5 ms^{-1}. Then

$$\text{time taken} \simeq \frac{120}{5} = 24 \text{ s},$$

which gives a total time of 34 s.

(If you have time to make a further check on the accuracy, you may like to go back and use a step size of $h = 0.1$s, or, instead of the Euler method, to use the modified Euler method for evaluating distance, or perhaps to use the step on part (d) of this SAQ.)

(d) A possible program is outlined below.

```
10 PRINT "PLEASE TYPE VALUES FOR A,B,C,D,E,F,G,H";
20 INPUT A,B,C,D,E,F,G,H
30 PRINT "INPUT INITIAL CONDITIONS FOR TIME
   AND VELOCITY";
40 INPUT X0,Y0
50 PRINT "INPUT STEP SIZE FOR TIME";
60 INPUT S
```

(These are the conditions required for the first order differential equation; we shall now incorporate the initial information required to calculate distance.)

```
70 PRINT "INPUT INITIAL HEIGHT";
80 INPUT H0
```

(We are now in a position to write the main framework or master segment of the program which will calculate time, velocity and distance, terminating the program when distance becomes negative.)

```
 90 PRINT "TIME", "VELOCITY", "HEIGHT"
100 PRINT X0,Y0,H0
110 H0 = H0 + S*Y0
120 GOSUB 5000
130 PRINT X0,Y0,H0
140 IF H0 < 0 THEN 9999
150 GOTO 110
```

(For the EUMSUB subroutine see Appendix 4.)

Selected values from the output are displayed below.

TIME	VELOCITY	HEIGHT
0	−50	200
.2	−35.8174	190
2	−6.02106	162.092
5.	−5.00349	146.446
10.	−5.	121.444
20.	−5.	71.4438
30.8001	−5.	17.4438
34.0001	−5.	1.44382
34.2001	−5.	.443817
34.4	−5.	−.556184

Solution to SAQ 7

B: The spring is calibrated in terms of the gravitational mass; it is not being used to compare different accelerations.

APPENDIX 1

The flow chart and the listing of the program to solve a first order differential equation by the Taylor series method are given below.

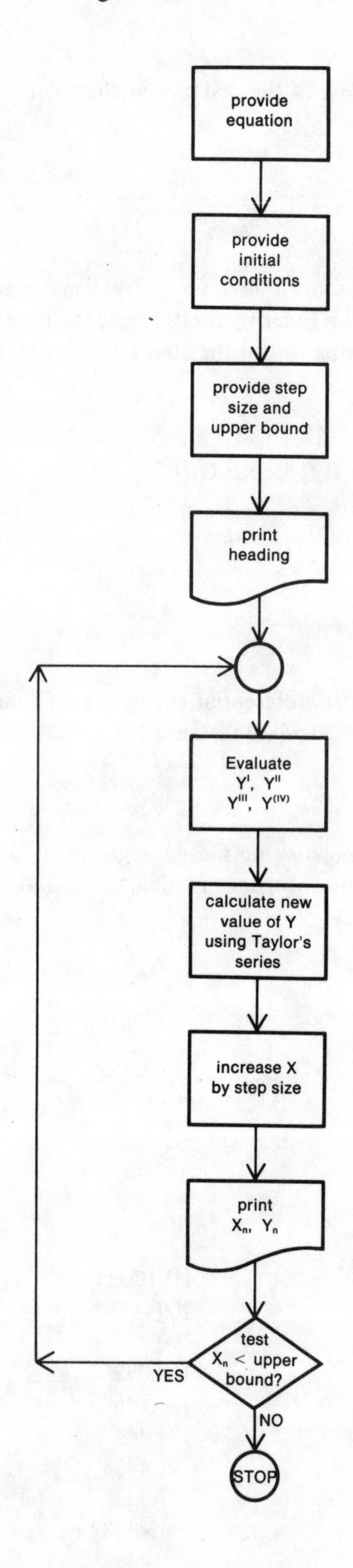

TYM282

```
10  PRINT "PROGRAM TO EVALUATE THE SOLUTION OF A
    DIFFERENTIAL"
20  PRINT "EQUATION NUMERICALLY USING TAYLOR'S SERIES"
30  PRINT "EQUATION IS IN THE FORM:—"
40  PRINT "             Y' = A*X ↑ B + C*Y ↑ D + E*(X ↑ F)*(Y ↑ G) + H"
50  PRINT "PLEASE TYPE VALUES OF A,B,C,D,E,F,G,H";
```

```
 60  INPUT A,B,C,D,E,F,G,H
 70  PRINT "PLEASE INPUT THE INITIAL CONDITIONS, I.E.
     VALUE OF"
 80  PRINT "Y FOR STARTING VALUE OF X"
 90  PRINT "TYPE VALUES FOR X AND Y";
100  INPUT X0,Y0
110  PRINT "PLEASE TYPE IN UPPER BOUND FOR RANGE OF X";
120  INPUT U
130  PRINT "PLEASE TYPE STEP SIZE FOR X";
140  INPUT S
150  PRINT
160  PRINT
170  PRINT TAB(25); "XN"; TAB(45); "YN"
180  PRINT
190  PRINT TAB(23);X0;TAB(43);Y0
200  IF X0 < > 0 THEN 220
210  X0 = 1.E - 10
220  IF Y0 < > 0 THEN 240
230  Y0 = 1.E - 10
240  Y1 = A*X0 ↑ B + C*Y0 ↑ D + E*(X0 ↑ F)*(Y0 ↑ G) + H
250  T1 = A*B*X0 ↑ (B - 1) + C*D*Y0 ↑ (D - 1)*Y1
260  T2 = (Y0 ↑ G)*F*X0 ↑ (F - 1) + X0 ↑ F*G*Y0 ↑ (G - 1)*Y1
270  Y2 = T1 + E*T2
280  T1 = A*B*(B - 1)*X0 ↑ (B - 2)
290  T2 = ((D - 1)*Y0 ↑ (D - 2)*Y1 ↑ 2 + Y0 ↑ (D - 1)*Y2)*C*D
300  T3 = E*F*((F - 1)*X0 ↑ (F - 2)*Y0 ↑ G + X0 ↑ (F - 1)*G*Y0 ↑ (G - 1)*Y1)
310  T4 = F*X0 ↑ (F - 1)*Y0 ↑ (G - 1)*Y1 + X0 ↑ F*(G - 1)*Y0 ↑ (G - 2)
     *Y1 ↑ 2 + X0 ↑ F*Y0 ↑ (G - 1)*Y2
320  Y3 = T1 + T2 + T3 + (T4*E*G)
330  T1 = A*B*(B - 1)*(B - 2)*X0 ↑ (B - 3)
340  T2 = C*D*(D - 1)*((D - 2)*Y0 ↑ (D - 3)*Y1 ↑ 3 + 2*Y0 ↑ (D - 2)*Y1*Y2)
350  T3 = C*D*((D - 1)*Y0 ↑ (D - 2)*Y1*Y2 + Y0 ↑ (D - 1)*Y3)
360  T4 = E*F*(F - 1)*((F - 2)*X0 ↑ (F - 3)*Y0 ↑ G + G*Y0 ↑ (G - 1)
     *Y1*X0 ↑ (F - 2))
370  T5 = (F - 1)*X0 ↑ (F - 2)*Y0 ↑ (G - 1)*Y1 + X0 ↑ (F - 1)*Y1 ↑ 2*
     (G - 1)*Y0 ↑ (G - 2)
380  T6 = 2*E*F*G*(X0 ↑ (F - 1)*Y0 ↑ (G - 1)*Y2)
390  T7 = F*X0 ↑ (F - 1)*Y0 ↑ (G - 2)*Y1 ↑ 2 + (G - 2)*Y0 ↑ (G - 3)*X0 ↑
     F*Y1 ↑ 3
400  T8 = E*G*(G - 1)*(X0 ↑ F*Y0 ↑ (G - 2)*2*Y1*Y2)
410  T9 = E*G*(F*X0 ↑ (F - 1)*Y0 ↑ (G - 1)*Y2 + (G - 1)*Y0 ↑ (G - 2)
     *X0 ↑ F*Y1*Y2)
420  T0 = E*G*X0 ↑ F*Y0 ↑ (G - 1)*Y3
430  Y4 = T1 + T2 + T3 + T4 + T5*(2*E*F*G) + T6 + T7*(E*G*(G - 1))
     + T8 + T9 + T0
440  Y0 = Y0 + S*Y1 + S ↑ 2/2*Y2 + S ↑ 3/6*Y3 + S ↑ 4/24*Y4
450  X0 = X0 + S
460  PRINT TAB(23);X0;TAB(43);Y0
470  IF X0 < U - .000001 THEN 240
480  STOP
490  END
```

(Note that:

(i) the TAB function which appears in the PRINT statement is used for printing spaces, e.g. PRINT TAB(25);X0 will result in the value of X0 being printed in the 26th column;

(ii) the symbol < > is used to denote "not equal to", i.e. X < > 0 means "X not equal to 0".)

APPENDIX 2

The flow chart and the listing of the program to solve a first order differential equation using Euler's method are given below.

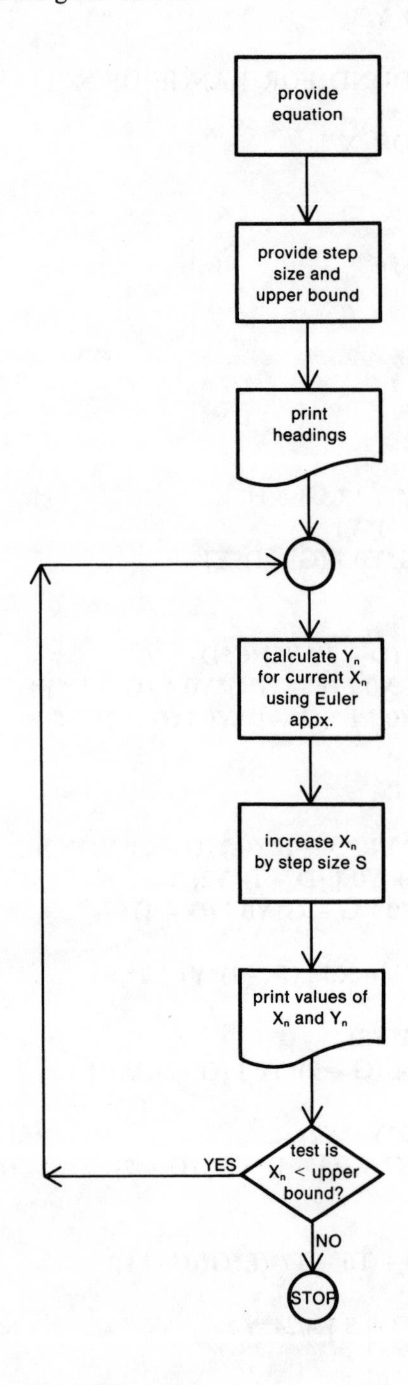

EUM282

```
 10  PRINT "PROGRAM TO EVALUATE THE SOLUTION OF A
     DIFFERENTIAL"
 20  PRINT "EQUATION NUMERICALLY USING EULER'S METHOD"
 30  PRINT "EQUATION IS IN THE FORM:—"
 40  PRINT "              Y' = A*X ↑ B + C*Y ↑ D + E*(X ↑ F)*(Y ↑ G) + H"
 50  PRINT "PLEASE TYPE VALUES FOR A,B,C,D,E,F,G,H";
 60  INPUT A,B,C,D,E,F,G,H
 70  PRINT "PLEASE INPUT THE INITIAL CONDITIONS,
     I.E. VALUE"
 80  PRINT "OF Y FOR STARTING VALUE OF X"
 90  PRINT "PLEASE TYPE VALUES FOR X AND Y";
100  INPUT X0,Y0
```

```
110 PRINT "PLEASE TYPE UPPER BOUND FOR RANGE OF X";
120 INPUT U
130 PRINT "PLEASE TYPE STEP SIZE FOR X";
140 INPUT S
150 PRINT "";TAB(25);"XN";TAB(45); "YN"
160 PRINT TAB(23);X0;TAB(43);Y0
170 IF X0 < > 0 THEN 190
180 X0 = 1.E - 10
190 IF Y0 < > 0 THEN 210
200 Y0 = 1.E - 10
210 Y0 = Y0 + S*(A*X0 ↑ B + C*Y0 ↑ D + E*X0 ↑ F*Y0 ↑ G + H)
220 X0 = X0 + S
230 PRINT TAB(23);X0;TAB(43);Y0
240 IF X0 < U - .000001 THEN 210
250 END
```

APPENDIX 3

The flow chart and the listing of a program to solve first order differential equations using the modified Euler method are given below.

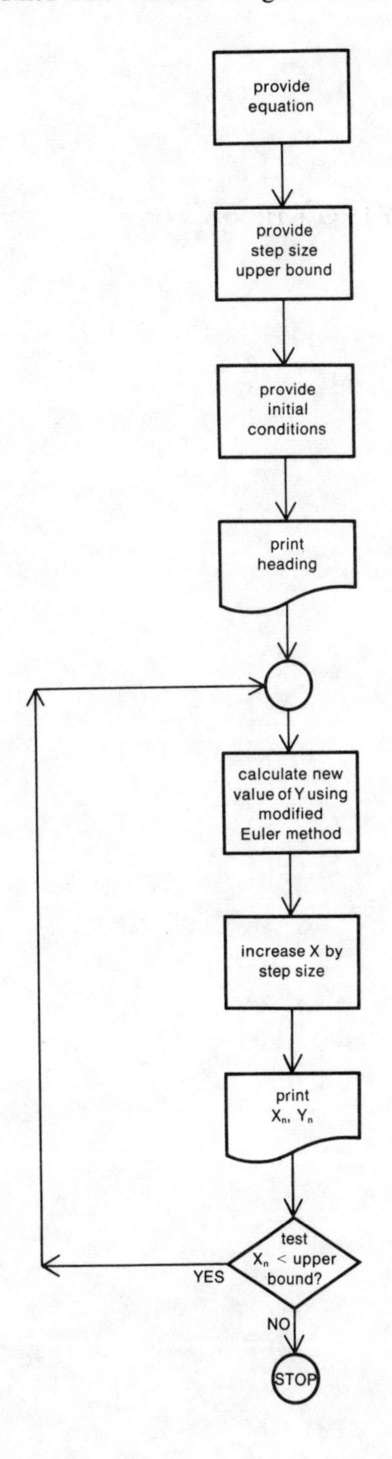

EMM282

```
10 PRINT "PROGRAM TO EVALUATE THE SOLUTION OF A
   DIFFERENTIAL"
20 PRINT "EQUATION NUMERICALLY USING A MODIFIED
   EULER METHOD"
30 PRINT "EQUATION IS IN THE FORM:—"
40 PRINT "              Y' = A*X ↑ B + C*Y ↑ D + E*(X ↑ F)*(Y ↑ G) + H"
50 PRINT "PLEASE TYPE VALUES FOR A,B,C,D,E,F,G,H";
60 INPUT A,B,C,D,E,F,G,H
70 PRINT "PLEASE INPUT THE INITIAL CONDITIONS,
   I.E. VALUE"
```

```
 80  PRINT "OF Y FOR STARTING VALUE OF X"
 90  PRINT "PLEASE TYPE VALUES FOR X AND Y";
100  INPUT X0,Y0
110  PRINT "PLEASE TYPE UPPER BOUND FOR RANGE OF X";
120  INPUT U
130  PRINT "PLEASE TYPE STEP SIZE FOR X";
140  INPUT S
150  PRINT "";TAB(25);"XN";TAB(45); "YN"
160  PRINT TAB(23);X0;TAB(43);Y0
170  IF X0 < > 0 THEN 190
180  X0 = 1.E - 10
190  IF Y0 < > 0 THEN 210
200  Y0 = 1.E - 10
210  Y3 = A*X0 ↑ B + C*Y0 ↑ D + E*X0 ↑ F*Y0 ↑ G + H
220  Y1 = Y0 + S*Y3
230  X1 = X0 + S
240  Y2 = A*X1 ↑ B + C*Y1 ↑ D + E*X1 ↑ F*Y1 ↑ G + H
250  Y1 = Y0 + S/2*(Y3 + Y2)
260  PRINT TAB(23);X1;TAB(43);Y1
270  Y0 = Y1
280  X0 = X1
290  IF X0 < U - .000001 THEN 210
300  STOP
310  END
```

APPENDIX 4

Program Names

EULSUB

EUMSUB

TAYSUB

Program Description

Although these programs exist as standard programs, their purpose is to provide subroutines which can be incorporated in any program which requires the solution of a first order differential equation of the form

$$Y' = AX \uparrow B + CY \uparrow D + E(X \uparrow F)(Y \uparrow G) + H$$

as part of its computation.

As the names suggest,

EULSUB employs the Euler method;
EUMSUB employs the modified Euler method;
TAYSUB employs the Taylor series expansion up to fourth order terms.

The values of A, B, C, D, E, F, G and H must be provided in the main body of the program. In addition, the current values of X and Y (in a sense, the initial conditions for the routine) must be set in locations X0 and Y0, and the step size for X is taken to be S.

On exit from the subroutine, X0 will have been incremented by S, and Y0 will contain the value of Y computed at the new value of X0 according to the method of the particular subroutine.

Operating Instructions

EULSUB uses statement numbers 5000–5016;
EUMSUB uses statement numbers 5000–5021;
TAYSUB uses statement numbers 5000–5036.

(There is an END statement at lines no. 9999.)

There are two methods of incorporating a subroutine from the library in a user's program.

(i) For the user not familiar with all the system commands, it is perhaps simplest to use

    GET—$ (appropriate subroutine name)

  Then type the main program statements and RUN the program. (Note: the LIST command will produce a properly ordered version of the complete program.)

(ii) The second method allows greater flexibility. Write the main body of the program and then simply add the subroutines by typing the command

    APPEND—$ (appropriate subroutine)

  If the Euler subroutine has been used, then the modified Euler method can be tested by using the commands

    DELETE—5000
    APPEND—$EUMSUB

  Then re-run the program.

When the subroutine is required by the main program, the introduction

n GOSUB 5000

is used, where n is the statement number.

Note

Check carefully the variables used in each subroutine, to avoid duplication in your main program (e.g. EUMSUB uses X1, Y1, Y2, Y3).

Program Listings

EULSUB

```
5000 REM .... SUBROUTINE TO SOLVE A DIFFERENTIAL EQUATION
     BY AN
5001 REM .... EULER METHOD. VALUES REQUIRED TO BE
     SUPPLIED BY
5002 REM .... THE MAIN PROGRAM ARE:— A,B,C,D,E,F,G,H,X0,
     Y0 AND S
5003 REM .... A,B,C,D,E,F,G,H ARE VALUES OF THE RESPECTIVE
     COEFFICIENTS
5004 REM .... IN THE DIFFERENTIAL EQUATION:—
5005 REM .... Y' = A*X ↑ B + C*Y ↑ D + E*X ↑ F*Y ↑ G + H
5006 REM .... X0 AND Y0 ARE THE INITIAL VALUES OF X AND Y,
     AND S IS THE
5007 REM .... STEP SIZE FOR X.
5008 REM .... THE VALUES RETURNED TO THE MAIN PROGRAM
     ARE:— THE VALUES
5009 REM .... OF X0 AND Y0 EVALUATED AT X0 + S
5010 IF X0 < > 0 THEN 5012
5011 X0 = 1.E - 10
5012 IF Y0 < > 0 THEN 5014
5013 Y0 = 1.E - 10
5014 Y0 = Y0 + S*(A*X0 ↑ B + C*Y0 ↑ D + E*X0 ↑ F*Y0 ↑ G + H)
5015 X0 = X0 + S
5016 RETURN
9999 END
```

TAYSUB

```
5000 REM .... SUBROUTINE TO SOLVE A DIFFERENTIAL EQUATION
     BY A
5001 REM .... TAYLOR'S SERIES. VALUES REQUIRED TO BE
     SUPPLIED BY
5002 REM .... THE MAIN PROGRAM ARE:— A,B,C,D,E,F,G,H,X0,
     Y0 AND S
5003 REM .... A,B,C,D,E,F,G,H ARE VALUES OF THE RESPECTIVE
     COEFFICIENTS
5004 REM .... IN THE DIFFERENTIAL EQUATION:—
5005 REM .... Y' = A*X ↑ B + C*Y ↑ D + E*X ↑ F*Y ↑ G + H
5006 REM .... X0 AND Y0 ARE THE INITIAL VALUES OF X AND Y,
     AND S IS THE
5007 REM .... STEP SIZE FOR X.
5008 REM .... THE VALUES RETURNED TO THE MAIN PROGRAM
     ARE:— THE VALUES
5009 REM .... OF X0 AND Y0 EVALUATED AT X0 + S
5010 IF X0 < > 0 THEN 5012
5011 X0 = 1.E - 10
5012 IF Y0 < > 0 THEN 5014
5013 Y0 = 1.E - 10
5014 Y1 = A*X0 ↑ B + C*Y0 ↑ D + E*(X0 ↑ F)*(Y0 ↑ G) + H
5015 T1 = A*B*X0 ↑ (B - 1) + C*D*Y0 ↑ (D - 1)*Y1
5016 T2 = (Y0 ↑ G)*F*X0 ↑ (F - 1) + X0 ↑ F*G*Y0 ↑ (G - 1)*Y1
5017 Y2 = T1 + E*T2
5018 T1 = A*B*(B - 1)*X0 ↑ (B - 2)
5019 T2 = ((D - 1)*Y0 ↑ (D - 2)*Y1 ↑ 2 + Y0 ↑ (D - 1)*Y2)*C*D
5020 T3 = E*F*((F - 1)*X0 ↑ (F - 2)*Y0 ↑ G + X0 ↑ (F - 1)*G*Y0 ↑ (G - 1)
     *Y1)
5021 T4 = F*X0 ↑ (F - 1)*Y0 ↑ (G - 1)*Y1 + X0 ↑ F*(G - 1)*Y0 ↑ (G - 2)
     *Y1 ↑ 2 + X0 ↑ F*Y0 ↑ (G - 1)*Y2
```

```
5022 Y3 = T1 + T2 + T3 + (T4*E*G)
5023 T1 = A*B*(B - 1)*(B - 2)*X0 ↑ (B - 3)
5024 T2 = C*D*(D - 1)*((D - 2)*Y0 ↑ (D - 3)*Y1 ↑ 3 + 2*Y0 ↑ (D - 2)*Y1*Y2)
5025 T3 = C*D*((D - 1)*Y0 ↑ (D - 2)*Y1*Y2 + Y0 ↑ (D - 1)*Y3)
5026 T4 = E*F*(F - 1)*((F - 2)*X0 ↑ (F - 3)*Y0 ↑ G + G*Y0 ↑ (G - 1)
     *Y1*X0 ↑ (F - 2))
5027 T5 = (F - 1)*X0 ↑ (F - 2)*Y0 ↑ (G - 1)*Y1 + X0 ↑ (F - 1)*Y1 ↑ 2*
     (G - 1)*Y0 ↑ (G - 2)
5028 T6 = 2*E*F*G*(X0 ↑ (F - 1)*Y0 ↑ (G - 1)*Y2)
5029 T7 = F*X0 ↑ (F - 1)*Y0 ↑ (G - 2)*Y1 ↑ 2 + (G - 2)*Y0 ↑ (G - 3)*X0 ↑
     F*Y1 ↑ 3
5030 T8 = E*G*(G - 1)*(X0 ↑ F*Y0 ↑ (G - 2)*2*Y1*Y2)
5031 T9 = E*G*(F*X0 ↑ (F - 1)*Y0 ↑ (G - 1)*Y2 + (G - 1)*Y0 ↑ (G - 2)
     *X0 ↑ F*Y1*Y2)
5032 T0 = E*G*X0 ↑ F*Y0 ↑ (G - 1)*Y3
5033 Y4 = T1 + T2 + T3 + T4 + T5*(2*E*F*G) + T6 + T7*(E*G*(G - 1))
     + T8 + T9 + T0
5034 Y0 = Y0 + S*Y1 + S ↑ 2/2*Y2 + S ↑ 3/6*Y3 + S ↑ 4/24*Y4
5035 X0 = X0 + S
5036 RETURN
9999 END
```

EUMSUB

```
5000 REM .... SUBROUTINE TO SOLVE A DIFFERENTIAL EQUATION
     BY A
5001 REM .... MODIFIED EULER METHOD. VALUES REQUIRED
     TO BE SUPPLIED BY
5002 REM .... THE MAIN PROGRAM ARE:— A,B,C,D,E,F,G,H,X0,
     Y0 AND S
5003 REM .... A,B,C,D,E,F,G,H ARE VALUES OF THE RESPECTIVE
     COEFFICIENTS
5004 REM .... IN THE DIFFERENTIAL EQUATION:—
5005 REM .... Y' = A*X ↑ B + C*Y ↑ D + E*X ↑ F*Y ↑ G + H
5006 REM .... X0 AND Y0 ARE THE INITIAL VALUES OF X AND Y,
     AND S IS THE
5007 REM .... STEP SIZE FOR X.
5008 REM .... THE VALUES RETURNED TO THE MAIN PROGRAM
     ARE:— THE VALUES
5009 REM .... OF X0 AND Y0 EVALUATED AT X0 + S
5010 IF X0 < > 0 THEN 5012
5011 X0 = 1.E - 10
5012 IF Y0 < > 0 THEN 5014
5013 Y0 = 1.E - 10
5014 Y3 = A*X0 ↑ B + C*Y0 ↑ D + E*X0 ↑ F*Y0 ↑ G + H
5015 Y1 = Y0 + S*Y3
5016 X1 = X0 + S
5017 Y2 = A*X1 ↑ B + C*Y1 ↑ D + E*X1 ↑ F*Y1 ↑ G + H
5018 Y1 = Y0 + S/2*(Y3 + Y2)
5019 Y0 = Y1
5020 X0 = X1
5021 RETURN
9999 END
```

APPENDIX 5

Additional Notes for Smith and Smith, *Mechanics*

In this appendix we are providing some additional explanatory notes for the parts of the set book we have asked you to read.

If you have found any parts difficult, you should turn to the appropriate page reference of S. These page and line references are on the left-hand side of the pages of this appendix. In some cases we indicate, for example, where proofs can be found in other books, and in other cases we add some further explanation of our own.

Page 60, line 2

Notice that the magnitude of $\mathbf{F}_A$ is

$$\gamma m_A m_B \frac{|\mathbf{r}_B - \mathbf{r}_A|}{|\mathbf{r}_B - \mathbf{r}_A|^3} = \frac{\gamma m_A m_B}{|\mathbf{r}_B - \mathbf{r}_A|^2},$$

which is why we call the law of gravitation an "inverse square law".

Page 62, line 2

"Uniform" means constant in this context.

Page 66, line −10

We are told that

(i) $\ddot{x} = -g \sin \beta$ and $\ddot{z} = -g \cos \beta$

and

$$x = z = 0, \quad \dot{x} = V \cos \alpha, \quad \dot{z} = V \sin \alpha \text{ at } t = 0.$$

Integrating both of equations (i), we obtain

(ii) $\dot{x} = -gt \sin \beta + A$ and $\dot{z} = -gt \cos \beta + B.$

Using the initial conditions, we find

$$A = V \cos \alpha \text{ and } B = V \sin \alpha$$

so

(iii) $\dot{x} = -gt \sin \beta + V \cos \alpha$ and $\dot{z} = -gt \cos \beta + V \sin \alpha.$

Integrating again, we obtain

$$x = -\frac{gt^2}{2} \sin \beta + Vt \cos \alpha + C,$$

$$z = -\frac{gt^2}{2} \cos \beta + Vt \sin \alpha + D.$$

Using the initial conditions $x = z = 0$, we obtain $C = D = 0$; hence

$$x = -\frac{gt^2}{2} \sin \beta + Vt \cos \alpha$$

and

$$z = -\frac{gt^2}{2} \cos \beta + Vt \sin \alpha.$$

Page 67, line 4

Using

$$\sin (A + B) = \sin A \cos B + \cos A \sin B$$

and

$$\sin (A - B) = \sin A \cos B - \cos A \sin B,$$

we obtain

$$\sin(A+B) - \sin(A-B) = 2\cos A \sin B.$$

(Trigonometric identities were discussed in *Refresher Booklet 10* (*RB 10*), available to Mathematics Foundation Course students.)

In line 4 there is the expression $2 \sin \alpha \cos(\alpha + \beta)$. The choice $A = \alpha + \beta$ and $B = \alpha$ produces the identity required.

Glossary

Terms defined in this glossary are printed in CAPITALS. Some terms not defined in this list may be found in the *Mathematical Handbook*.

		Page
FIELD OF FORCE	If at every point of a region of space a particle is acted on by a force, then there is said to exist a FIELD OF FORCE in that region. The force will, in general, depend on the position and possibly the time. Mathematically, A FIELD OF FORCE on a subset U of (3-dimensional) space is a function which assigns to each point $x \in U$ the force acting at x. Thus it may be considered as a function from U to the product of U with the set of all force vectors, assigning to each $x \in U$ the pair $(x, \mathbf{F})$. (The ordered pair then denotes a force $\mathbf{F}$ acting at x.)	9
GRAVITATIONAL CONSTANT	The GRAVITATIONAL CONSTANT is the factor of proportionality occurring in NEWTON'S LAW OF GRAVITATION.	9
GRAVITATIONAL FORCE	See NEWTON'S LAW OF GRAVITATION.	9
GRAVITATIONAL MASS	The GRAVITATIONAL MASS of a body is its mass determined by using NEWTON'S LAW OF GRAVITATION.	23
INERTIAL MASS	The INERTIAL MASS, M_I, of a body is the ratio $\frac{F}{a}$, where $\mathbf{F}$ is the force applied to the body and $\mathbf{a}$ is its acceleration.	23
NEWTON'S LAW OF GRAVITATION	NEWTON'S LAW OF GRAVITATION states: any two particles of masses m_1, m_2, distance r apart, attract each other with a force $\gamma \frac{m_1 m_2}{r^2}$ (where γ is the GRAVITATIONAL CONSTANT). This attraction is called a GRAVITATIONAL FORCE.	9
NUMERICAL METHODS:		
(i) TAYLOR SERIES METHOD	This method uses TAYLOR'S SERIES $y(x_0 + h) = y(x_0) + hy'(x_0) + \frac{h^2}{2!} y''(x_0) + \ldots$ to find solutions to a differential equation in tabular form. The more terms are used the greater the accuracy of the solution. When n terms of the series are used, the method amounts to approximating the solution by a polynomial of degree $(n-1)$.	12
(ii) EULER'S METHOD	This method depends on the assumption that a curve may be approximated in a small interval by its tangent at one end-point of the interval.	14

(iii) PREDICTOR-CORRECTOR METHODS	These methods use two formulas. The PREDICTOR formula is used to make an estimate of a new value of the solution. The CORRECTOR formula is then used on this estimate and the previous value to give a better estimate.	15
TERMINAL SPEED	The TERMINAL SPEED is the speed at which a body moves in a medium when the resultant force on the body is zero. For a projectile falling vertically, the terminal speed is the speed when the air resistance balances gravity.	10
UNIFORM FIELD OF FORCE	A UNIFORM FIELD OF FORCE on a subset U of (3-dimensional) space is a constant FIELD OF FORCE; i.e. it is a function which assigns to each $x \in U$ the pair $(x, \mathbf{F})$ where the vector $\mathbf{F}$ is the same for each $x \in U$.	9

Unit 6 Rigid Bodies

Contents

Page

Bibliography

T. M. Apostol, *Calculus Vol. II* (Blaisdell, 1969).

A. H. Lightstone, *Concepts of Calculus II* (Harper, 1966).

These books give a rigorous approach to volume and surface integrals. You may find them too difficult at a first reading, but, if you wish to extend yourself mathematically, they are books to which you can refer.

W. Kaplan, *Advanced Calculus* (Addison-Wesley, 1962).

M. R. Spiegel, *Vector Analysis*, Shaum Outline Series (McGraw-Hill, 1959).

You will find these books useful if you require further examples and exercises on volume and surface integrals.

Note

References to the Open University Mathematics Foundation Course Units (The Open University Press, 1971) take the form *Unit M100 3, Operations and Morphisms*.

Objectives

After working through this unit you should be able to:

(i) describe what is meant by a *rigid body* and define its mass-centre, linear momentum and angular (or moment of) momentum;
(ii) define the various forces which a rigid body may experience, and distinguish between them;
(iii) solve simple problems involving rigid bodies similar to those in this text;
(iv) define surface and volume integrals and evaluate such integrals;
(v) define the moment of inertia of a lamina about some axis and evaluate it in simple cases;
(vi) define the centre of gravity of a rigid body;
(vii) define the centre of pressure of a rigid body when immersed in a fluid;
(viii) discuss the basic principles of the Gravity Dam;
(ix) define impulse and understand what is meant by a perfectly elastic collision.

Study Sequence

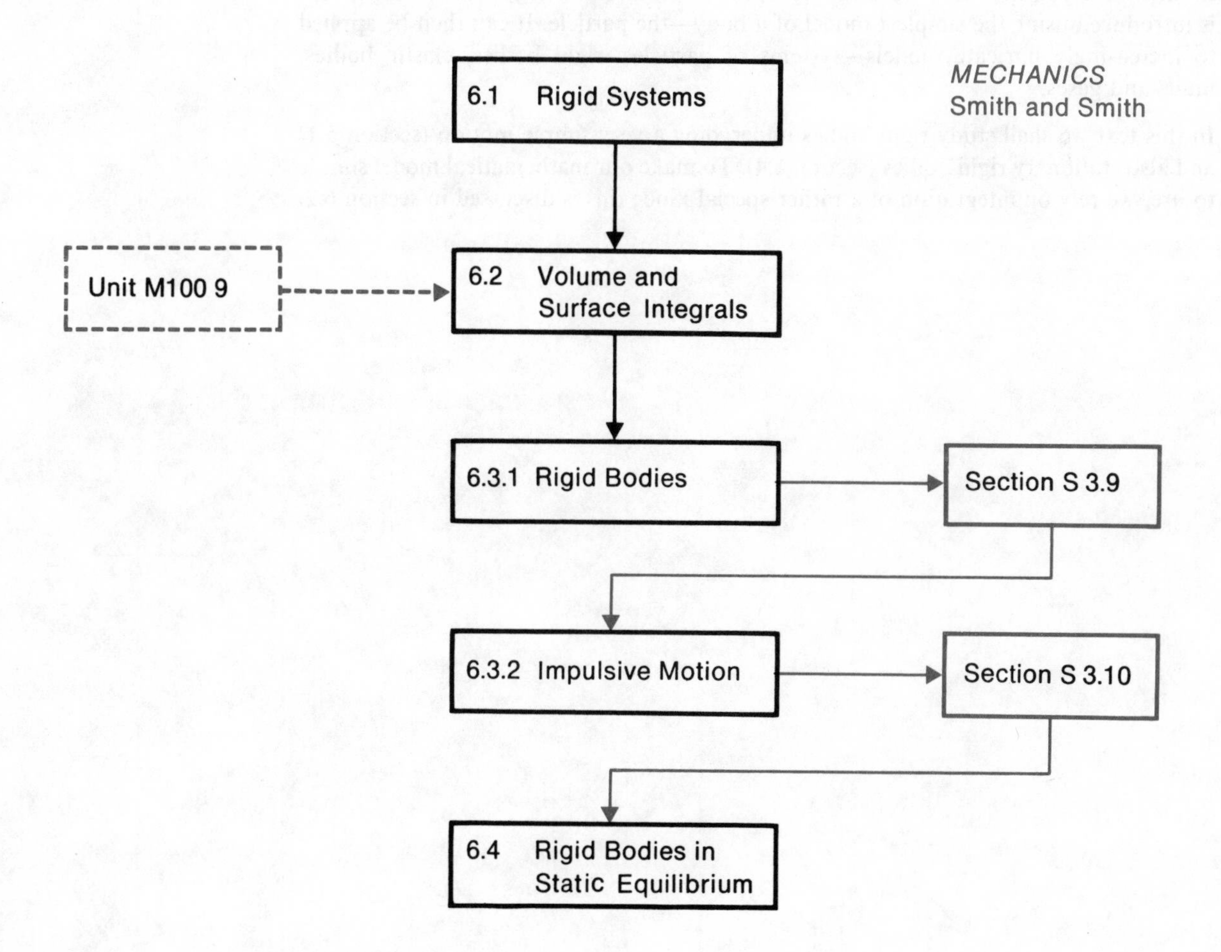

6.0 INTRODUCTION

As this course proceeds we are developing the fundamental concepts of mechanics. These include concepts such as force, linear momentum and angular momentum. In later units you will meet other concepts such as work and energy. Each new concept is introduced using the simplest model of a body—the particle. It can then be applied to increasingly intricate models—systems of particles, rigid bodies, elastic bodies, fluids and gases.

In this text we shall study rigid bodies undergoing a very simple motion (section 6.1) and also stationary rigid bodies (section 6.4). To make our mathematical model simple to use, we rely on integration of a rather special kind; this is discussed in section 6.2.

6.1 RIGID SYSTEMS

In *Unit 3* we used Newton's laws to study first the motion of a single particle, and then a system of particles acted on by forces. Two points emerged from this study.

First, we found that it is useful to define two vector quantities:

(i) the linear momentum of a particle of mass m, $\mathbf{p} = m\mathbf{v}$, where $\mathbf{v}$ is the velocity of the particle;

(ii) the angular momentum of the particle, $\mathbf{h} = m\mathbf{r} \times \mathbf{v}$, where $\mathbf{r}$ is the position vector of the particle relative to a *fixed* origin.

These two quantities are important because they remain invariant in common physical situations—the linear momentum when no forces are acting on the particle, the angular momentum when the moment of the forces about the fixed origin is zero.

The second point is that we can examine one important aspect of the behaviour of a system of n particles by modelling the system with a single particle, whose mass is the total mass of the system situated at a point defined as the centre of mass. This equivalent particle then behaves as if all the external forces acting on the particles of the system were acting on it. To study the motions of the *individual* particles still requires the solution of n differential equations

$$\mathbf{F}_i = m_i \ddot{\mathbf{r}}_i \qquad (i = 1, 2, \ldots, n),$$

which, although they are simple to write down, could be very complicated to solve. However, we can obtain an important clue to the *total* behaviour of the system by concentrating our attention on the centre of mass.

To develop a "rigid body" from a system of particles requires two steps. Firstly, we incorporate the restriction on the system that "the distances between every pair of particles of the system remain fixed": we then have a *rigid system*. Secondly, we assume that the matter in a body is continuously distributed. To understand this, consider the following limiting process. Inside a given fixed volume (the volume of the body), we imagine a rigid system. This system is replaced by another of the same mass and volume, in which the number of particles is increased (and thus the spacing between the particles is decreased). This system in its turn is replaced by another in which there are yet more particles. The limit of this sequence is a *rigid body*. The rigid body of mechanics is as idealized as the particle of mechanics. This is because real bodies are composed of atoms which cannot be subdivided indefinitely, and the separation of the atoms does not remain fixed.

What difference does it make to our analysis of the motion of a system of particles if we make the system rigid? We approach this question by considering a particular example. We choose the example of the motion of a pendulum. It is not the usual kind of pendulum with a weight on the end of a string, but a solid steel bar of length $2a$ swinging on a knife edge as in (i) in the figure.

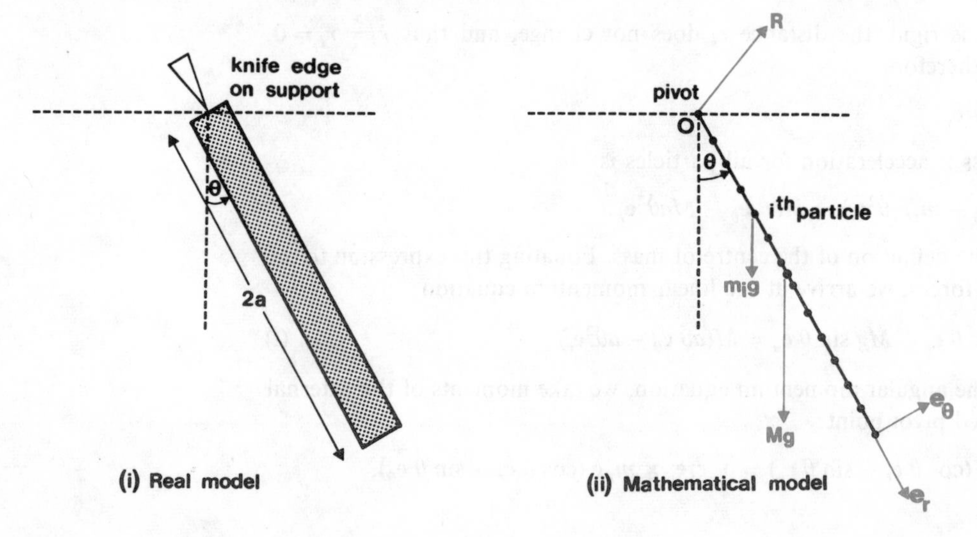

For our mathematical model, (ii) in the figure, we make two assumptions: first, that the knife edge is the pivot at the very end of the bar, and second, that the width of the bar is negligible compared with its length. As a consequence of the latter assumption, the bar is replaced by a string of particles rigidly fixed relative to each other. The two quantities to determine are **R**, the reaction at the pivot, and θ; these are variables dependent on the time t. The two equations to use are the linear momentum equation and the angular momentum equation.

SAQ 1

Indicate the word or phrase which completes the following sentence correctly.

The linear momentum equation for a system of particles is:

sum of _________ forces = total mass × acceleration of centre of mass

A external
B internal
C internal and external.

(Solution is given on p. 33.)

SAQ 2

Indicate the phrase which completes the following sentence correctly.
The angular momentum equation for a system of particles is:
moment of external forces about a *fixed point O* = _________

A sum of angular momenta of particles about O
B rate of change of sum of angular momenta of particles about O
C rate of change of angular momentum of total mass at mass-centre about O.

(Solution is given on p. 33.)

The external force on a particle of mass m_i is $m_i g$ acting vertically downwards. Since we are dealing with motion in a circle, it is convenient to use the unit vectors $\mathbf{e}_r$ and $\mathbf{e}_\theta$. The force on the particle is then

$$m_i g \cos\theta\, \mathbf{e}_r - m_i g \sin\theta\, \mathbf{e}_\theta .$$

Thus the sum of the external forces is

$$\begin{aligned}\mathbf{R} + \sum_i (m_i g \cos\theta\, \mathbf{e}_r - m_i g \sin\theta\, \mathbf{e}_\theta) &= \mathbf{R} + g\cos\theta\, \mathbf{e}_r \sum_i m_i - g \sin\theta\, \mathbf{e}_\theta \sum_i m_i \\ &= \mathbf{R} + Mg\cos\theta\, \mathbf{e}_r - Mg\sin\theta\, \mathbf{e}_\theta ,\end{aligned}$$

where M is the total mass of the bar.

We look next at the accelerations involved. The acceleration of the ith particle is (from p. S19)

$$(\ddot{r}_i - r_i\dot{\theta}^2)\, \mathbf{e}_r + (2\dot{r}_i\dot{\theta} + r_i\ddot{\theta})\, \mathbf{e}_\theta .$$

Because the system is rigid, the distance r_i does not change, and thus $\dot{r}_i = \ddot{r}_i = 0$. The acceleration is therefore

$$r_i\ddot{\theta}\, \mathbf{e}_\theta - r_i\dot{\theta}^2 \mathbf{e}_r .$$

The sum of the mass × acceleration for all particles is

$$\sum_i (m_i r_i \ddot{\theta}\, \mathbf{e}_\theta - m_i r_i \dot{\theta}^2 \mathbf{e}_r) = Ma\ddot{\theta}\, \mathbf{e}_\theta - Ma\dot{\theta}^2 \mathbf{e}_r .$$

This follows from the definition of the centre of mass. Equating this expression to the sum of the external forces, we arrive at the linear momentum equation:

$$\mathbf{R} + Mg\cos\theta\, \mathbf{e}_r - Mg\sin\theta\, \mathbf{e}_\theta = M(a\ddot{\theta}\, \mathbf{e}_\theta - a\dot{\theta}^2 \mathbf{e}_r). \qquad \text{(i)}$$

In order to obtain the angular momentum equation, we take moments of the external forces about the fixed pivot point:

$$\sum_i \mathbf{r}_i \times m_i g (\cos\theta\, \mathbf{e}_r - \sin\theta\, \mathbf{e}_\theta) = \sum_i r_i \mathbf{e}_r \times m_i g (\cos\theta\, \mathbf{e}_r - \sin\theta\, \mathbf{e}_\theta).$$

SAQ 3

Calculate:

(i) $\mathbf{e}_r \times \mathbf{e}_r$
(ii) $\mathbf{e}_r \times \mathbf{e}_\theta$.

(Solution is given on p. 33.)

Using the solution to SAQ 3, we have for the moment of the external forces

$$-g \sin \theta \, \mathbf{k} \sum_i m_i r_i = -Mga \sin \theta \, \mathbf{k},$$

where $\mathbf{k}$ is a unit vector coming out of the page in the figure. The rate of change of angular momentum about the pivot point, O, is $\frac{d}{dt}(\sum_i \mathbf{r}_i \times m_i \mathbf{v}_i)$, where $\mathbf{v}_i$ is the velocity of the ith particle. Because the radial velocity, i.e., $\dot{\mathbf{r}}_i$, is zero, we can write the rate of change of angular momentum as

$$\frac{d}{dt}\left(\sum_i \mathbf{r}_i \times m_i (r_i \dot{\theta} \, \mathbf{e}_\theta)\right) = \frac{d}{dt}\left(\sum_i m_i r_i^2 \dot{\theta} \mathbf{k}\right) = \ddot{\theta}\mathbf{k} \sum_i m_i r_i^2.$$

The angular momentum equation for the system of particles is therefore

$$-Mga \sin \theta \mathbf{k} = \ddot{\theta}\mathbf{k} \sum_i m_i r_i^2$$

or

$$-Mga \sin \theta = I_o \ddot{\theta}, \qquad \text{(ii)}$$

where $I_o = \sum_i m_i r_i^2$.

The quantity I_o is a physical property of the bar and is known as its moment of inertia about the point O. It is a physical property in the same sense that mass is a physical property. The parallel is a close one, for I_o appears in equation (ii) in much the same manner as M appears in equation (i). The mass is the relevant physical quantity for translational motion; the moment of inertia is the corresponding quantity for rotational motion. If we could determine I_o, we could find θ from equation (ii) and then use equation (i) to find $\mathbf{R}$. This would solve the problem.

Clearly, as we improve the model by using more and more particles, we shall encounter sums of the form $\sum_i m_i r_i^2$ which will be impossible to evaluate for purely practical reasons. The solution is to simplify our mathematical model and to introduce the concept of a rigid body in which matter is taken to be continuously distributed. We are then able to replace the cumbersome summations over many particles by a sophisticated and elegant integration procedure. Before we can do this we must extend our knowledge of integration.

Summary

In this section we have examined a particular example of a rigid system of particles, namely a model of a swinging bar. When we derived the equations of motion of such a system from the linear and angular momentum equations, we found it useful to define a quantity, known as the *moment of inertia*, which plays a similar part in rotational motion to that played by mass in translational motion.

6.2 VOLUME AND SURFACE INTEGRALS

6.2.0 Introduction

(Television programme 5 is closely associated with a considerable part of the material of this section.)

In this section we shall first remind you of the basic mathematical ideas of the definite integral. We shall concentrate on those particular aspects needed in order to define the volume and surface integrals (section 6.2.1). We then define the volume integral and look at the methods of calculating it (section 6.2.2). Finally we adopt the same approach to the surface integral (section 6.2.3).

In our study of volume and surface integrals we shall use primarily intuitive arguments. For a more rigorous approach you could consult T. M. Apostol, *Calculus, Vol. II* or A. H. Lightstone, *Concepts of Calculus II.* (See Bibliography.)

6.2.1 The Definite Integral

In this section we revise the definition of a definite integral; we then carry forward the basic ideas to new situations.

Geometrically, if we take the graph of some positive function $f(x)$ (depicted in the figure), which is continuous over an interval $[a, b]$ in its domain, then the definite integral of f over $[a, b]$, $\int_a^b f(x)\,dx$, is the area beneath the graph of f between a and b.

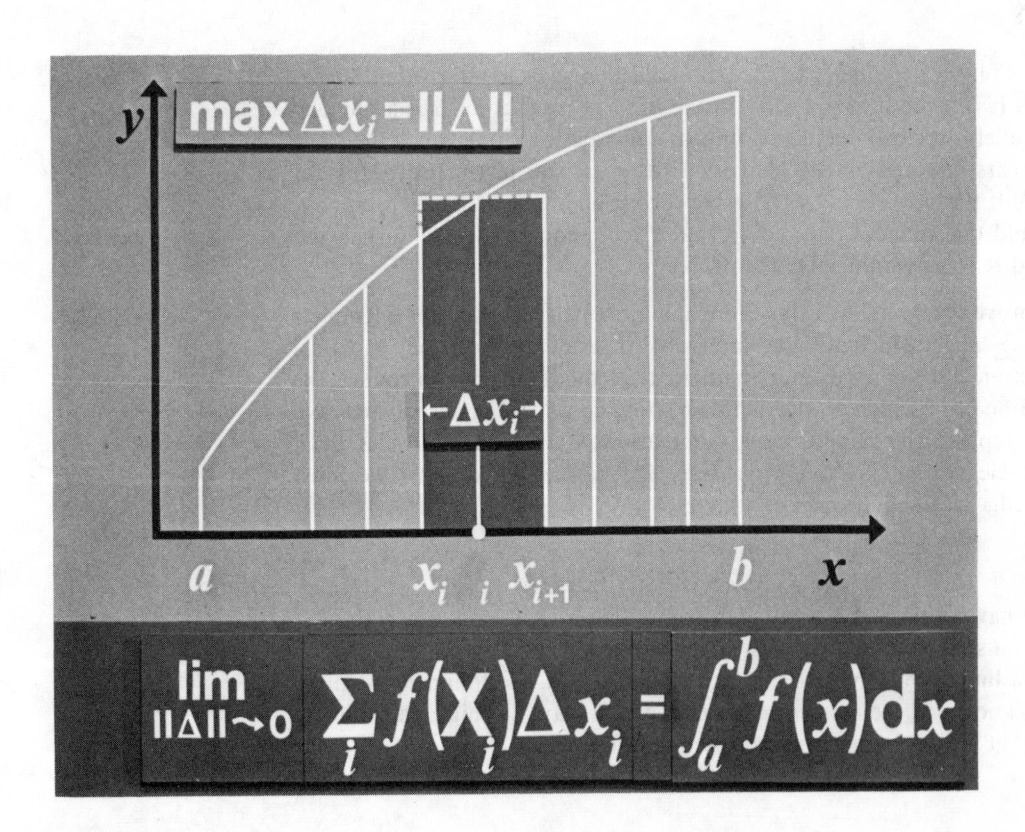

SAQ 4

Is it always true that the area beneath the graph of an arbitrary function f between a and b is $\int_a^b f(x)\,dx$? Give a counter-example if it is not.

(Solution is given on p. 33.)

We define this area and the definite integral in the following way. The interval $[a, b]$ is subdivided into intervals of widths $\Delta x_i = x_{i+1} - x_i$, which may be of equal width (as in *Unit M100 9, Integration I*) but need not be. If they are not of equal width, we label the maximum Δx_i as $\|\Delta\|$. We take any point P_i in $[x_i, x_{i+1}]$ with co-ordinate X_i and construct the rectangle of height $f(X_i)$ on the base of width Δx_i. The sum of the areas of all the rectangles

$$\sum_i f(X_i)\,\Delta x_i$$

is then an approximation to the area under the graph. We then take the limit of the sum as $\|\Delta\|$ tends to zero ($\|\Delta\|$ to ensure that *all* the sub-intervals tend to zero) and define this limit, if it exists, to be $\int_a^b f(x)\,dx$, i.e.

$$\lim_{\|\Delta\| \to 0} \sum_i f(X_i)\Delta x_i = \int_a^b f(x)\,dx.$$

In sections 6.2.2 and 6.2.3, we shall use similar expressions to define volume and surface integrals.

6.2.2 The Volume Integral

Suppose that f is a continuous† function having the set of all points in 3-dimensional space as its domain. The interval $[a, b]$ in the 1-dimensional case is replaced by the defined volume V in the 3-dimensional case, and the intervals Δx_i are replaced by small volumes (of any shape provided they fit together), ΔV_i, enclosing points P_i. These volumes fill the entire volume V.

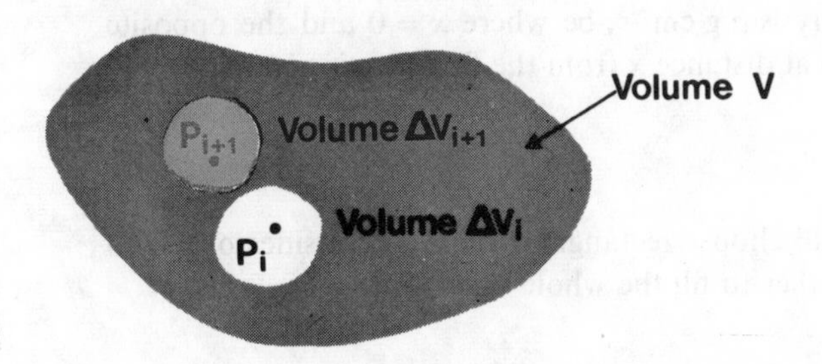

We now form the products $f(P_i)\,\Delta V_i$, then form the sum $\sum_i f(P_i)\,\Delta V_i$, and take the limit

$$\lim_{\|\Delta\| \to 0} \sum_i f(P_i)\,\Delta V_i,$$

where $\|\Delta\|$ is now the maximum width* of the volumes ΔV_i for any partitioning. If this limit exists, it is written as

$$\int_V f(P)\,dV$$

and is called the volume integral of the function f over the volume V. The following example demonstrates the principle.

† We have not formally defined continuity of a function of more than one variable. The function f of three variables x, y, z is continuous at the point (a, b, c) if $f(x, y, z)$ approaches $f(a, b, c)$ as x approaches a, y approaches b and z approaches c in any manner.

* Normally the word "width" is replaced in standard texts by the phrase "diameter of the set" which is defined to be the maximum distance between any two points in the set.

Example

In a cube of width 10 cm the density (defined as mass per unit volume) changes linearly from a g cm^{-3} at one face to b g cm^{-3} at the opposite face. What is the mass of the cube?

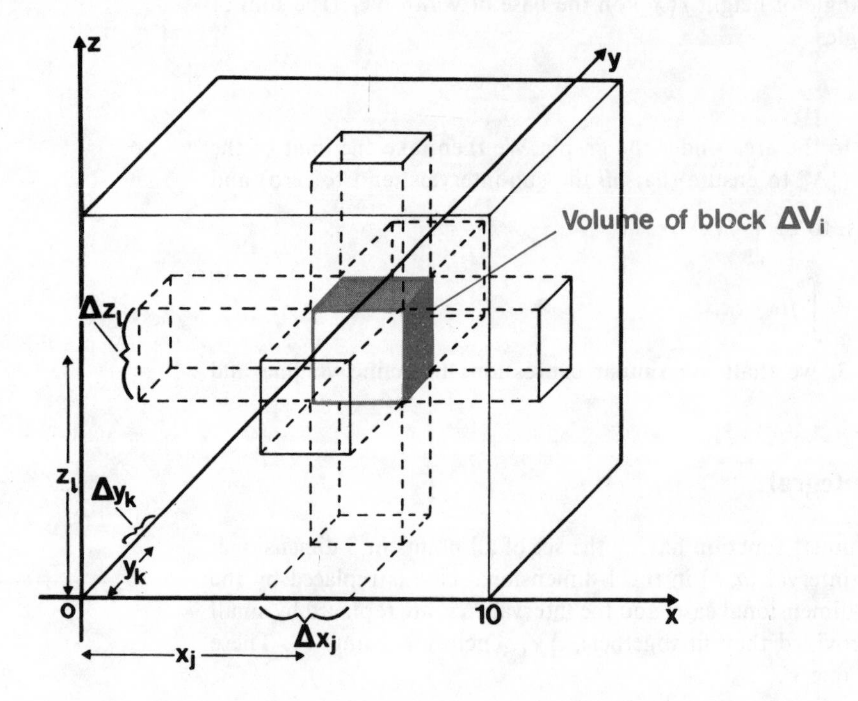

Let the first face, at which the density is a g cm^{-3}, be where $x = 0$ and the opposite face be where $x = 10$. The density ρ at distance x from the first face is given by

$$\rho = \left(a + \left(\frac{b-a}{10}\right)x\right).$$

For the small volumes, ΔV, we could choose rectangular blocks with sides of length Δx, Δy and Δz which would fit together to fill the whole of V. Then

$$\Delta V = \Delta x\, \Delta y\, \Delta z.$$

The density ρ_i could be calculated at, say, the centre of each small cube and the products $\rho_i \Delta V_i$ formed and summed over the large cube. The notation begins to look cumbersome, but we obtain an approximation to the total mass:

$$\sum_i \rho_i\, \Delta V_i = \sum_j \sum_k \sum_l \rho_{jkl}\, \Delta x_j\, \Delta y_k\, \Delta z_l,$$

where the volume ΔV_i is centred at (x_j, y_k, z_l) and has sides Δx_j, Δy_k, and Δz_l. The density, ρ_i, is that appropriate to the position (x_j, y_k, z_l), which we have denoted by ρ_{jkl}.

If we used small cubes of width 1 cm as building blocks, the summation over i would be from 1 to 1000 and the summations over j, k, l from 1 to 10. This would clearly be a tedious method of calculating the mass, and it can be avoided if we represent the volume integral in terms of the definite integrals we already know. This involves two analytic steps which we shall not prove, but which are intuitively fairly obvious.

The first step is to assume that

$$\lim_{\|\Delta\| \to 0} \sum_i \rho_i\, \Delta V_i =$$

$$\lim_{\|\Delta_l\| \to 0} \sum_l \left(\lim_{\|\Delta_k\| \to 0} \sum_k \left(\lim_{\|\Delta_j\| \to 0} \sum_j \rho_{jkl}\, \Delta x_j \right) \Delta y_k \right) \Delta z_l,$$

where $\|\Delta_l\| =$ maximum Δz_l, and so on. Geometrically, this implies that we can take the limit in three stages. First we add all the rectangular blocks in lines parallel to the x-axis and form that limit. Then we add all the lines of boxes along the y-axis and take that limit. Finally, we pile the flat boxes vertically to make the cube and take that limit.

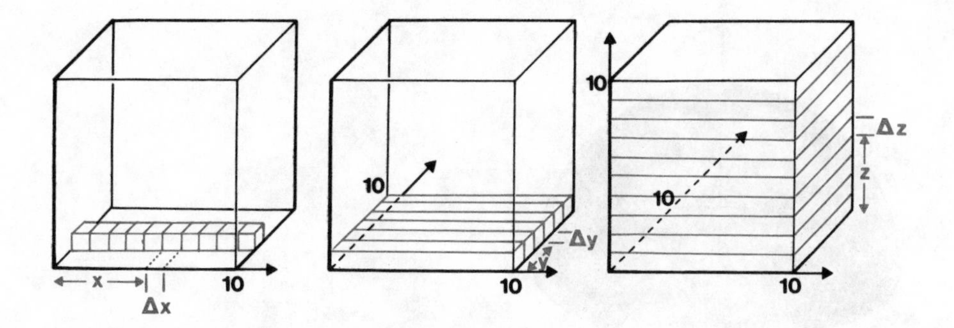

The second step is to assume that each separate limit is an appropriate definite integral, so that

$$\int_V \rho \, dV = \int_0^{10} \boxed{\int_0^{10} \boxed{\int_0^{10} \rho \, dx} \, dy} \, dz.$$

We actually calculate the volume integral by calculating the definite integrals one by one—the innermost one first. Integrals such as the above are called *repeated integrals* or *triple integrals* or *multiple integrals*. In this example we have

$$\int_0^{10} \rho \, dx = \int_0^{10} \left(a + \frac{b-a}{10} x\right) dx = \left[ax + \frac{(b-a)}{20} x^2\right]_0^{10}$$
$$= 5(a+b),$$

$$\int_0^{10} \boxed{\int_0^{10} \rho \, dx} \, dy = \int_0^{10} 5(a+b) \, dy = 50(a+b),$$

and

$$\int_0^{10} \boxed{\int_0^{10} \boxed{\int_0^{10} \rho \, dx} \, dy} \, dz = \int_0^{10} 50(a+b) \, dz = 500(a+b),$$

the units being grams. Normally the volume integral is written

$$\int_0^{10} \int_0^{10} \int_0^{10} \rho \, dx \, dy \, dz.$$

The order in which the integration is to be carried out has to be inferred from the order in which the variables appear. In this example, the order was unimportant. The following SAQ illustrates an example where this is not the case. It shows that it is wise to choose a co-ordinate system appropriate to the problem. Having chosen a co-ordinate system, we must then obtain the correct expression for the basic volume in terms of these co-ordinates.

SAQ 5

Confirm that the volume of the hemisphere of radius a is $\dfrac{2\pi a^3}{3}$ by writing down the volume integral for the hemisphere in cylindrical polar co-ordinates R, θ, z.

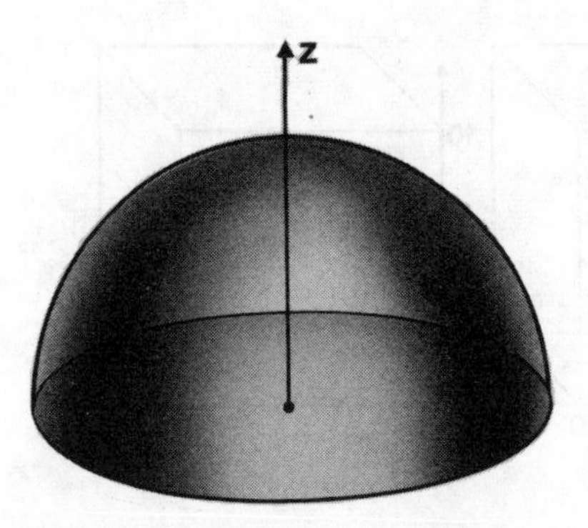

(Solution is given on p. 33.)

In this section so far we have been finding volume integrals of scalar functions of position. The same principle is used in finding volume integrals for vector functions of position: we return to the basic summations. Then we obtain, for a given vector function $\mathbf{f}(P)$,

$$\int_V \mathbf{f}(P)\, dV.$$

We calculate this integral by choosing an appropriate co-ordinate system and calculating the scalar components of the vector. Such an integral is required, for example, if the gravitational attraction of a large body is calculated by integrating, or adding up, the effects of all the particles of which it is composed.

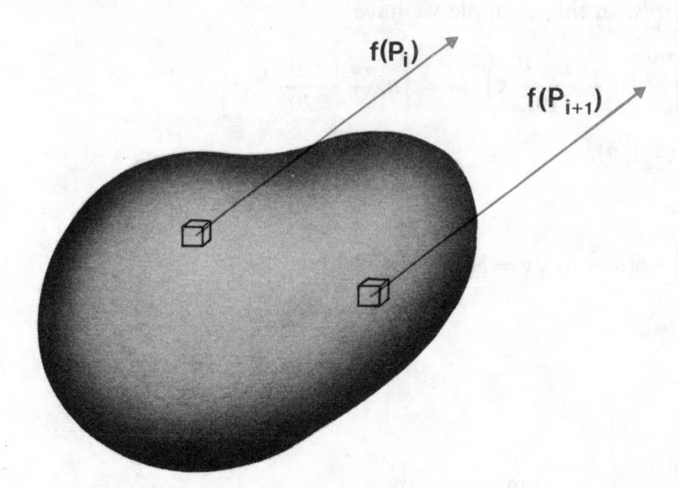

Such an integration was performed in television programme 5 and is repeated below as an example.

Example

Evaluate $\int_V \mathbf{f}(P)\, dV$, where V is a hemisphere of radius a, P is the position shown in the diagram, and $\mathbf{f}(P) = \mathbf{r}$, the position vector of a point.

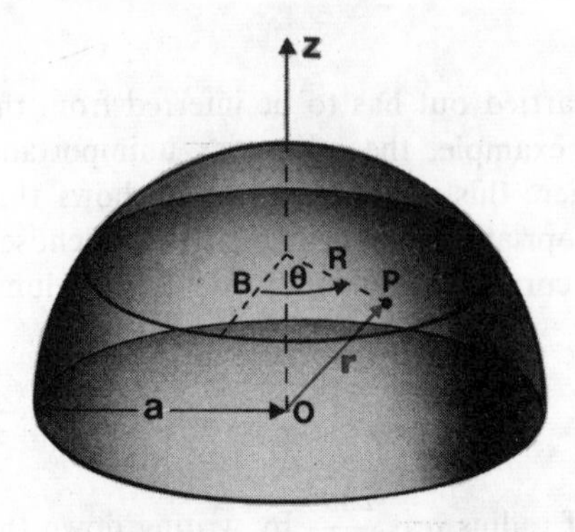

Transforming the volume integral to a triple integral and determining the limits of integration of the definite integrals follows the same procedure as that used in SAQ 5.

In this example we obtain the triple integral

$$\int_0^a \left[\int_0^B \left[\int_0^{2\pi} \mathbf{r}\, d\theta \right] R\, dR \right] dz.$$

In the television programme, we use symmetry to show that this is

$$\int_0^a \left[\int_0^B \left[\int_0^{2\pi} \mathbf{k} z\, d\theta \right] R\, dR \right] dz$$

but here we shall express $\mathbf{r}$ in Cartesian components,

$$\mathbf{r} = R\cos\theta\, \mathbf{i} + R\sin\theta\, \mathbf{j} + z\mathbf{k}.$$

(At first sight it might appear that we should put $\mathbf{r} = R\mathbf{e}_R + z\mathbf{k}$, using cylindrical polar components, but then $\mathbf{e}_R$ would be dependent on θ.)

Then

$$\int_0^{2\pi} (R\cos\theta\, \mathbf{i} + R\sin\theta\, \mathbf{j} + z\mathbf{k})\, d\theta$$

$$= R\mathbf{i}\Big[\sin\theta\Big]_0^{2\pi} + R\mathbf{j}\Big[-\cos\theta\Big]_0^{2\pi} + 2\pi\, z\mathbf{k}$$

$$= 2\pi\, z\mathbf{k}.$$

The second integral then becomes

$$\mathbf{k}\int_0^B 2\pi z\, R\, dR = \pi z B^2 \mathbf{k},$$

and finally the third integral becomes

$$\mathbf{k}\int_0^a \pi z B^2\, dz = \mathbf{k}\int_0^a \pi z(a^2 - z^2)\, dz$$

$$= \left(\pi\frac{a^4}{2} - \pi\frac{a^4}{4}\right)\mathbf{k}$$

$$= \pi\frac{a^4}{4}\,\mathbf{k}.$$

SAQ 6

Evaluate $\int_V \mathbf{r}\, dV$, where V is a cone with height h and vertical angle α in the position shown in the diagram. (We shall see integrals of this form later in the text when we consider centres of mass.)

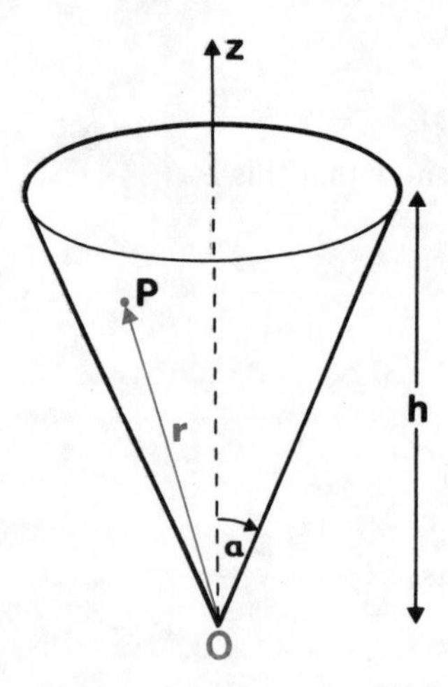

(Solution is given on p. 34.)

6.2.3 The Surface Integral

The extension of the idea of the limit of a sum to *surface integrals* and the manipulation of the surface integral into a double integral are simple in theory.

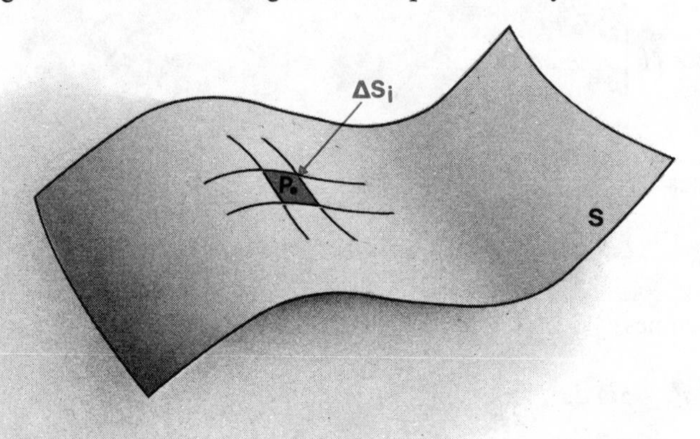

If f is a continuous function having the set of all points in 3-dimensional space as its domain, and S is a surface, then we can define the surface integral

$$\int_S f(S)\, dS \quad \text{to be} \quad \lim_{\|\Delta\| \to 0} \sum_i f(P_i)\, \Delta S_i\,,$$

provided this limit exists, where $\|\Delta\|$ is now the maximum "width" of the ΔS_i at each subdivision. In this case, "width" is defined as follows. For every two points on the boundary of the surface element there is, on the surface, a path of least length joining them. The "width" is the maximum length of all such paths for the various pairs of points.

Surface Integrals on a Single Plane

We choose the plane to be the xy-plane, and use A to represent the flat surface area. In calculating $\int_A f(P)\, dA$ the most difficult part is the determination of the limits of integration in the equivalent double integral, so we shall concentrate first on that aspect. To make the notation clearer, whenever we write a *double* integral we shall put the one to be performed first in red.

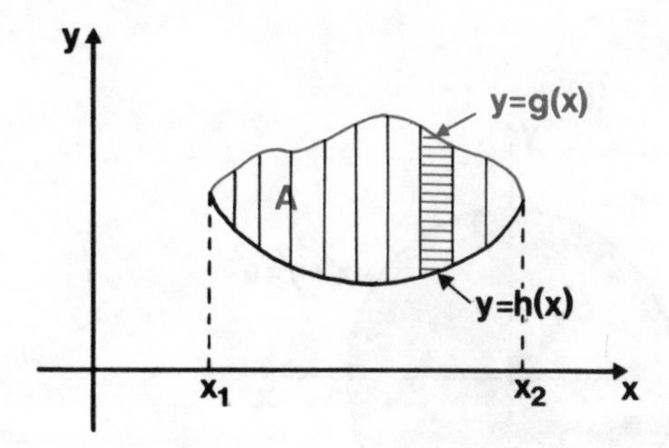

Consider the area A in the figure. If we sum the rectangles vertically first, and then sum the vertical strips horizontally, we obtain

$$\sum_i f(P_i)\,\Delta A_i = \sum_j \left(\sum_k F(x, y)\,\Delta y_k \right) \Delta x_j,$$

the k summation being taken from $y = h(x)$ to $y = g(x)$, and the j summation being taken from $x = x_1$ to $x = x_2$. It follows from the definition that

$$\int_A f(P)\,dA = \int_{x_1}^{x_2} \int_{h(x)}^{g(x)} F(x, y)\,dy\,dx.$$

Example

Evaluate $\int_A xy\,dA$, where A is the area shown in the figure.

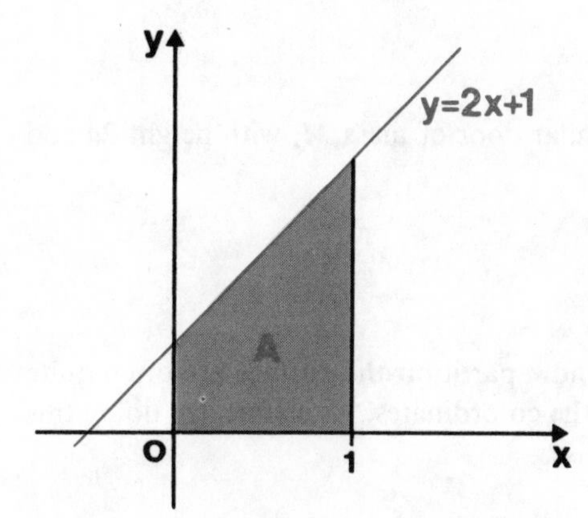

In this case,

$$\int_A xy\,dA = \int_0^1 \int_0^{2x+1} xy\,dy\,dx.$$

The limits on the red integral are derived first. The calculation is then straightforward.

$$\int_0^1 \int_0^{2x+1} xy\,dy\,dx = \int_0^1 x \left[\frac{y^2}{2} \right]_0^{2x+1} dx$$

$$= \int_0^1 \frac{x(2x+1)^2}{2}\,dx$$

$$= \frac{1}{2} \left[x^4 + \frac{4x^3}{3} + \frac{x^2}{2} \right]_0^1 = \frac{17}{12}.$$

(It is not essential to integrate over y and use vertical strips first—we could integrate over x first, using horizontal strips. In the present case we would then obtain a sum of two double integrals.)

SAQ 7

Fill in the limits of the definite integrals for the double integrals appropriate to each of the following figures.

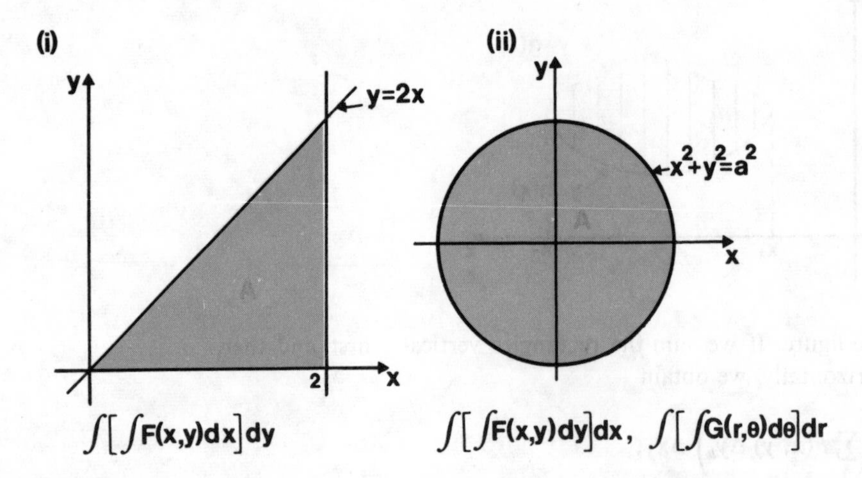

(Solution is given on p. 35.)

In section 6.1 we introduced a new quantity, $\sum_i m_i r_i^2$, for a system of particles, which we defined as its *moment of inertia*. (m_i is the mass of a particle at a distance r_i from the axis about which the body rotates.) If we model a plane lamina of uniform mass per unit area, ρ, by such a system, then the moment of inertia is $\sum_i \rho r_i^2 \, \Delta A$.

Taking the limit produces a uniform lamina with moment of inertia

$$\int_A \rho r^2 \, dA.$$

SAQ 8

Evaluate the moment of inertia of a rectangular door of mass M, with height $2a$ and width a, about its hinges.

(Solution is given on p. 35.)

Surface Integrals on a Curved Surface

The small surface elements into which we now partition the surface are often quite difficult to visualize and express in terms of the co-ordinates. As a start, try doing this for the simple surface in the following exercise.

SAQ 9

Confirm that the surface area of a sphere of radius a is $4\pi a^2$ by evaluating $\int_S dS$ over the surface, using as co-ordinates the "longitude", ϕ, and "latitude", θ, of a point.

(Solution is given on p. 36.)

The next example is a problem which we discuss in television programme 5; here we tackle it in a different way to emphasize the choice that can occur in calculations.

Example

Evaluate $\int_S \mathbf{r}\, dS$, where S is the surface shown in the figure.

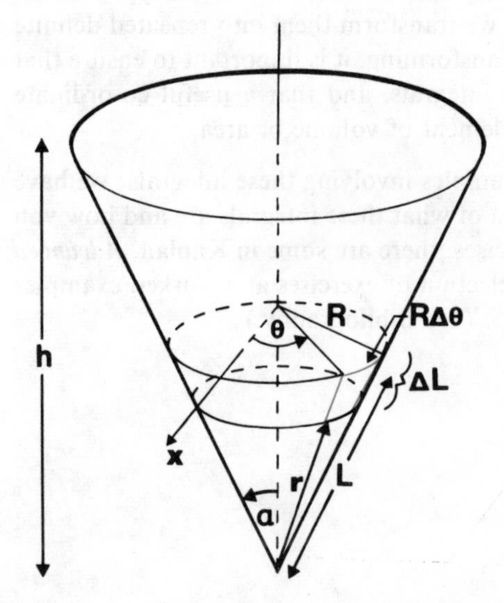

The surface element has area $R\, \Delta\theta\, \Delta L$ to the approximation we require. (If you wish to evaluate this area exactly, open out the cone by imagining it to be made of a piece of paper, and find the areas of the appropriate small sectors of annuli so produced.)

In the television programme we replace the variable L by the variable z, but the integral can be evaluated just as easily using the variable L. Then we obtain

$$\int_S \mathbf{r}\, dS = \int_0^{h/\cos\alpha} \int_0^{2\pi} \mathbf{r}\, R\, d\theta\, dL.$$

Since $R = L \sin\alpha$, $z = L\cos\alpha$, the integral becomes

$$\int_0^{h/\cos\alpha} \int_0^{2\pi} (L\sin\alpha\cos\theta\,\mathbf{i} + L\sin\alpha\sin\theta\,\mathbf{j} + L\cos\alpha\mathbf{k})L\sin\alpha\, d\theta\, dL$$

$$= \mathbf{k}\int_0^{h/\cos\alpha} 2\pi L^2 \cos\alpha\sin\alpha\, dL.$$

(The other two components of the vector are zero. Remember that L is constant as θ varies.) Thus

$$\int_S \mathbf{r}\, dS = \frac{2\pi h^3 \sin\alpha}{3\cos^2\alpha}\mathbf{k}.$$

SAQ 10

Evaluate $\int_S \mathbf{r}\, dS$ over the surface of the hemisphere shown.

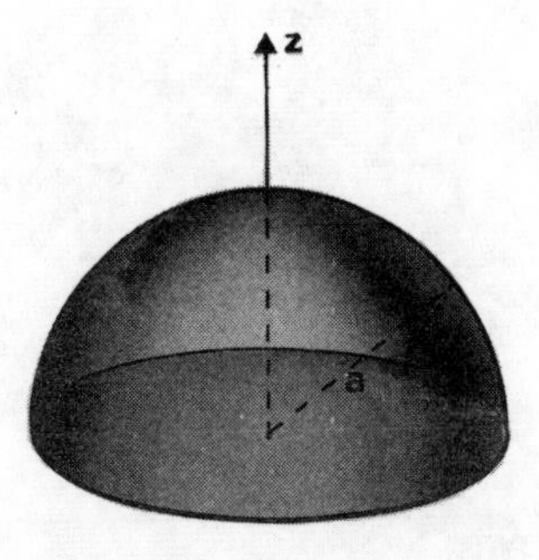

(Solution is given on p. 37.)

6.2.4 Summary

In this section we have demonstrated how surface and volume integrals are based on the same mathematical idea as the definite integral—the limit of a certain type of sum. When we use these integrals in calculations, we transform them into repeated definite integrals which can be handled directly. In transforming, it is important to ensure that we obtain the correct limits of the definite integrals, and that a useful co-ordinate system is chosen for determining the basic element of volume or area.

Much time can be spent on exercises and examples involving these integrals; we have provided a small selection to give you an idea of what these integrals are and how you can calculate them. If you want further exercises, there are some in Kaplan, *Advanced Calculus*, section 5-10, and there is a good selection of exercises and worked examples in M. R. Spiegel, *Vector Analysis*, Chapter 5. (See Bibliography.)

6.3 "READ THE BOOK" SECTION

6.3.1 Rigid Bodies

We now use the set book to bring together the ideas of a rigid system, volume and surface integrals, and the definition of a rigid body. The basic equations governing the motion of such a body, and the linear momentum and angular momentum (or moment of momentum) equations, are extensions of the equations previously derived for systems of particles.

Read section S3.9, p. S69 up to the end of the solution to Example 10 on p. S73.

SAQ 11

Exercise 7, p. S86.

(It is not necessary to do this exercise from first principles.)

(Solution is given on p. 37.)

SAQ 12

Find the centre of mass of the uniform L-shaped lamina shown in the following diagram.

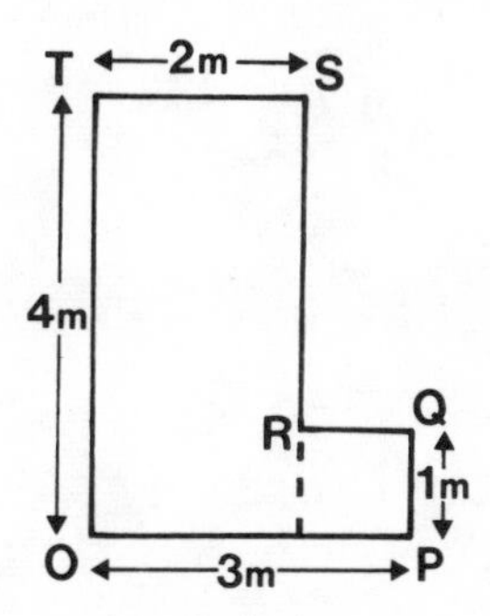

(Solution is given on p. 38.)

SAQ 13

Find the centre of mass of a uniform right circular cone of height h and base radius a.

(Solution is given on p. 39.)

Read section S3.9 from p. S73 to the end of the section. (You may find this difficult to follow at the first reading.)

Note

On p. S76, the linear momentum and angular momentum equations are listed (Equation (21)). For the latter equation, moments are taken about a *fixed* point. In fact the equation still holds if moments are taken about the centre of mass, regardless of how it is moving or accelerating. This statement requires proof and we have incorporated this in Appendix 1 (p. 46). From now on in problems you may take moments *either* about a fixed point *or* about the centre of mass.

SAQ 14

Exercise 14, p. S86.

(Solution is given on p. 40.)

SAQ 15

Exercise 15, p. S87.

(Solution is given on p. 40.)

6.3.2 Impulsive Motion

Collisions are discussed in the final reading section of this text. You will find that to model collisions, or impulsive motions, mathematically requires the introduction of an empirical law governing the behaviour of certain bodies; it is called *Newton's law of restitution.*

Read section S3.10, pp. S78–85.

SAQ 16

Newton's law of restitution provides us with a simple mathematical model with which we are able to examine a possibly complex situation. Can you think of a situation in which this particular mathematical model is entirely inappropriate?

(Solution is given on p. 41.)

SAQ 17

Exercise 32, p. S89.

(Solution is given on p. 41.)

SAQ 18

Exercise 29, p. S89.

(Solution is given on p. 41.)

6.3.3 Summary

In this section, we have developed the basic mathematical tools for analysing rigid bodies; some of the definitions introduced are outlined below.

The position vector of the centre of mass, $\bar{\mathbf{r}}$, is defined by

$$M\bar{\mathbf{r}} = \int_V \mathbf{r}\rho \, dV,$$

where M is the mass of a rigid body of volume V.

For a rigid body,

$$\mathbf{F} = M\ddot{\bar{\mathbf{r}}}.$$

In other words, when forces are applied to a rigid body, the motion produced is the same as that which would arise if the resultant force were to act on a single particle of mass M situated at the centre of mass. Also

$$\mathbf{M} = \frac{d\mathbf{h}}{dt},$$

where $\mathbf{M}$ is the vector moment of the forces about a fixed point, or the centre of mass, and

$$\mathbf{h} = \int_V \mathbf{r} \times \dot{\mathbf{r}}\rho \, dV.$$

(N.B. scalar M is the mass: vector $\mathbf{M}$ is the moment.)

The linear momentum given to a body of mass m when its velocity is increased from $\mathbf{v}_1$ to $\mathbf{v}_2$ is defined as the *impulse* $\mathbf{I}$; that is,

$$\mathbf{I} = m(\mathbf{v}_2 - \mathbf{v}_1).$$

Newton's law of restitution states:

When two bodies collide, the component of their relative velocity after impact in the direction of the common normal at the point of impact is $(-e)$ times their relative approach velocity in this direction.

The coefficient of restitution, e, is positive and may be assumed to be constant for particular materials in collision. It has a value between zero and unity. If $e = 1$, then the bodies are said to be *perfectly elastic*.

6.4 RIGID BODIES IN STATIC EQUILIBRIUM

6.4.0 Introduction

In section S3.9, the equations of static equilibrium are derived:

$$\mathbf{F} = \mathbf{0}, \mathbf{M} = \mathbf{0}. \qquad \text{(p. S76)}$$

Example 12, p. S77 illustrates the use of these equations. Physical situations governed by these two equations give rise to a subject area known as Statics (as distinct from Dynamics, which is our main concern in this course). The predominant interest in Statics is in structures, for example, the frameworks of buildings or the design of a crane. We do not intend to delve far into that subject area in this course. However, in section 6.4.3 we shall make a case study of one particular structure—a gravity dam (and we shall also return to the subject in *Unit 9*). But before we do this, we introduce the concept of the centre of gravity of a rigid body (section 6.4.1). When rigid bodies are immersed in a fluid there is also another point of interest—the centre of pressure. We consider this in section 6.4.2.

6.4.1 Centre of Gravity

Suppose you take an irregular piece of cardboard, suspend it from a pin at a point A, and draw a vertical line AB through A by using a plumb-line. Suppose you then choose another point C, repeat the experiment, and draw another straight line CD.

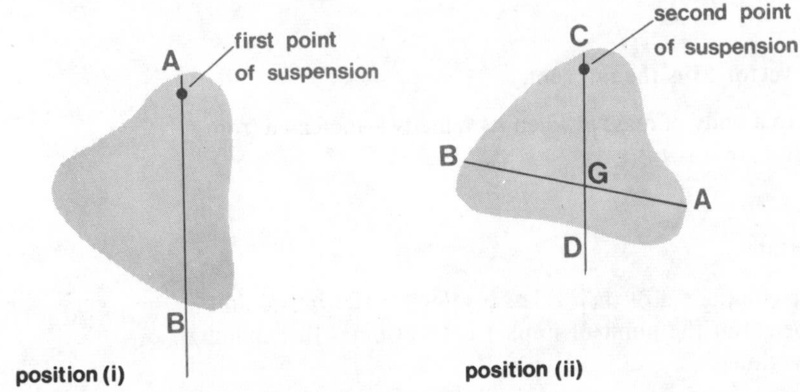

The point G of intersection of AB and CD is the *centre of gravity* of the piece of cardboard. It is the point through which the resultant gravitational force on the body may be supposed to act, *whatever* the orientation of the body. (We have yet to prove this.) In fact we shall show that the centre of gravity always coincides with the centre of mass of a rigid body in a *uniform* gravitational field. If the gravitational field is not uniform, the centre of gravity is not defined, as you will see in an SAQ later in this section.

Example

Show that the centre of gravity of a body coincides with its centre of mass if the gravitational field is uniform (that is, if the field is constant in magnitude and direction throughout the volume being considered).

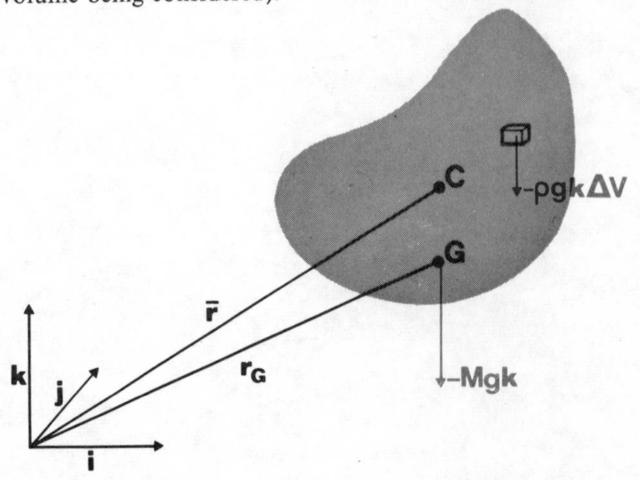

In a uniform gravitational field the vector force on each element of volume is $-\rho g\mathbf{k}\,\Delta V$ if $\mathbf{k}$ is pointing vertically upwards, and g is the acceleration due to gravity. The resultant vector force on the body will be

$$\int_V -\rho g\mathbf{k}\, dV = -\left(\int_V \rho\, dV\right) g\mathbf{k} = -Mg\mathbf{k},$$

where M is the mass of the body.

Along which line does the force act? To determine this, we use the equations for static equilibrium, namely

$$\mathbf{F} = \mathbf{0},\ \mathbf{M} = \mathbf{0}.$$

Theoretically, the rigid body will be in equilibrium if we introduce a force $Mg\mathbf{k}$ upwards through its centre of gravity G. Then the total moment about any fixed point, say the origin, will be zero. Thus

$$\mathbf{r}_G \times Mg\mathbf{k} + \int_V \mathbf{r} \times -\rho g\mathbf{k}\, dV = \mathbf{0},$$

where $\mathbf{r}_G$ is the position vector of the centre of gravity. That is,

$$\mathbf{r}_G \times Mg\mathbf{k} - g\left(\int_V \mathbf{r}\rho\, dV\right) \times \mathbf{k} = \mathbf{0}.$$

But, by definition, the position vector of the centre of mass is given by

$$\bar{\mathbf{r}} = \frac{1}{M}\int_V \mathbf{r}\rho\, dV.$$

Therefore

$$Mg\mathbf{r}_G \times \mathbf{k} - Mg\bar{\mathbf{r}} \times \mathbf{k} = \mathbf{0},$$

so

$$(\mathbf{r}_G - \bar{\mathbf{r}}) \times \mathbf{k} = \mathbf{0}.$$

SAQ 19

If $\mathbf{a} \times \mathbf{b} = \mathbf{0}$, what can we deduce about $\mathbf{a}$ and $\mathbf{b}$?

(Solution is given on p. 42.)

Using the result of the last SAQ, we deduce that either $\mathbf{r}_G = \bar{\mathbf{r}}$, or $(\mathbf{r}_G - \bar{\mathbf{r}})$ is parallel to $\mathbf{k}$, which implies that G and the centre of mass lie on the same vertical line.

The above analysis holds for any orientation of the body. The centre of mass and the centre of gravity are fixed points in the body, so the only way that they can always lie on the same vertical line is if they are *coincident*.

SAQ 20

The dumb-bell in the figure is rotated in the gravitational field shown. Use an intuitive argument to indicate whether or not the line of action of the resultant force on the dumb-bell will always cross the line AB at the same point.

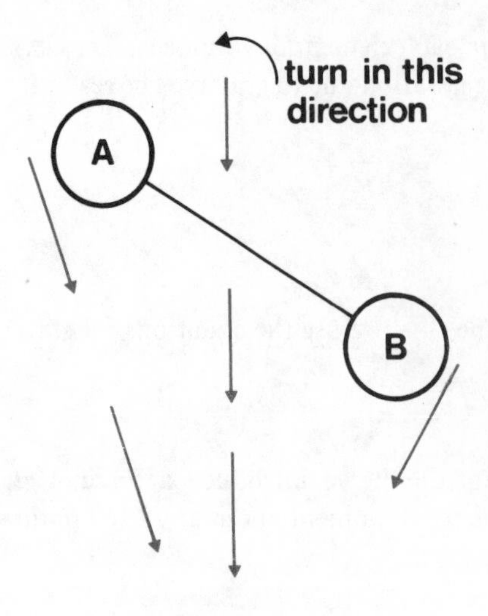

(Solution is given on p. 42.)

From the last SAQ it becomes clear that there is no unique point through which the force would always pass. In practice, therefore, the centre of gravity is a useful concept only in the context of a uniform gravitational field. It is then that the centre of gravity and the centre of mass are synonymous terms (although their derivation is different) and the calculation of the position of a centre of gravity is simply the calculation of a centre of mass. So there is an easy practical way of determining the centre of mass of an irregularly shaped body—one suspends it from different points and determines the intersection of the various vertical lines through those points.

Conversely, in determining the centre of gravity of a dam, say, we can determine its centre of mass.

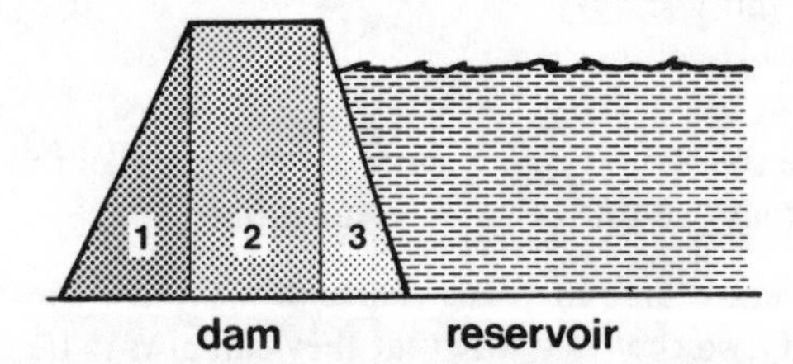

This is done by first calculating the centres of mass of each of the regularly shaped parts. Then the centre of mass of the composite structure is found using the equation

$$\bar{\mathbf{r}} = \frac{m_1\bar{\mathbf{r}}_1 + m_2\bar{\mathbf{r}}_2 + m_3\bar{\mathbf{r}}_3 + \cdots + m_n\bar{\mathbf{r}}_n}{m_1 + m_2 + m_3 + \cdots + m_n},$$

where m_1 and $\bar{\mathbf{r}}_1$ are the mass and position vector of the centre of mass of the first part, and so on.

SAQ 21

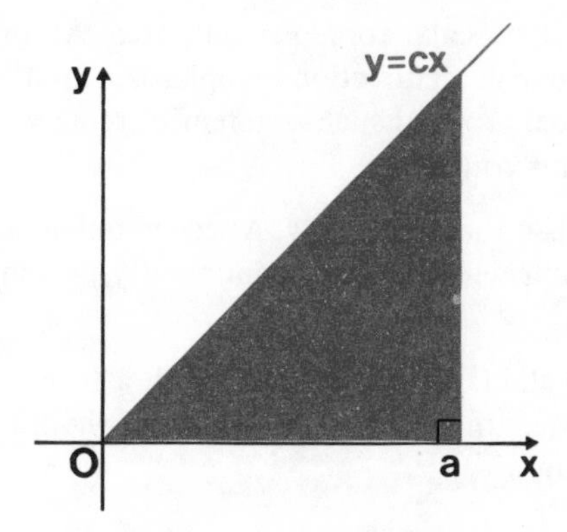

As a preliminary to the next SAQ, determine the position vector of the centre of mass of the triangle shown above.

(Solution is given on p. 42.)

SAQ 22

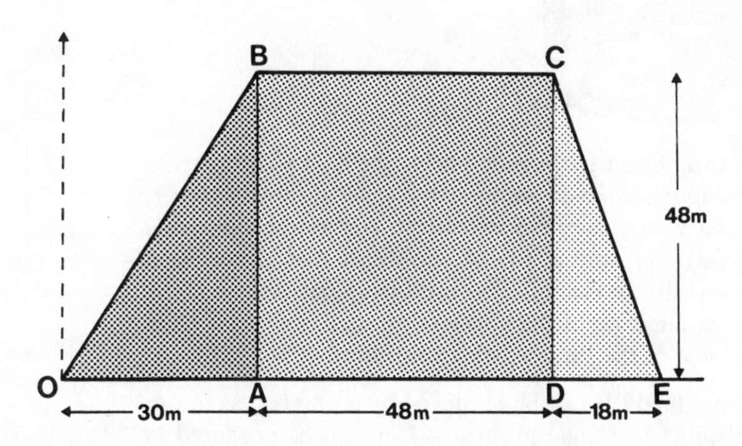

Determine the position of the centre of mass of the dam shown above, assuming that the material is of uniform density and the thickness of the dam is uniform perpendicular to the plane of the paper.

(Solution is given on p. 43.)

Summary

The centre of gravity of a rigid body is defined as the point through which the resultant gravitational force on the body acts, whatever the body's orientation. We have shown that, in a uniform gravitational field, the centre of gravity is coincident with the centre of mass: in a non-uniform field it is not uniquely defined.

6.4.2 Centre of Pressure

Pressure is a physical quantity which is defined as scalar force per unit area. At any point in a stationary field, the pressure is the same in all directions, emphasizing that it is a *scalar*, not a vector, quantity. The theoretical proof of such a statement requires a study of fluids which is beyond the scope of this course.

One obvious difference between solids and fluids is that fluids flow. A second difference is that solids can sustain forces acting at a point whereas fluids cannot—try pushing your finger into ice and then into water!

If a plunger is pushed into a tube containing fluid, as in the figure, with a force of magnitude F, the resulting effect is transmitted throughout the whole of the fluid almost instantaneously by means of the pressure.

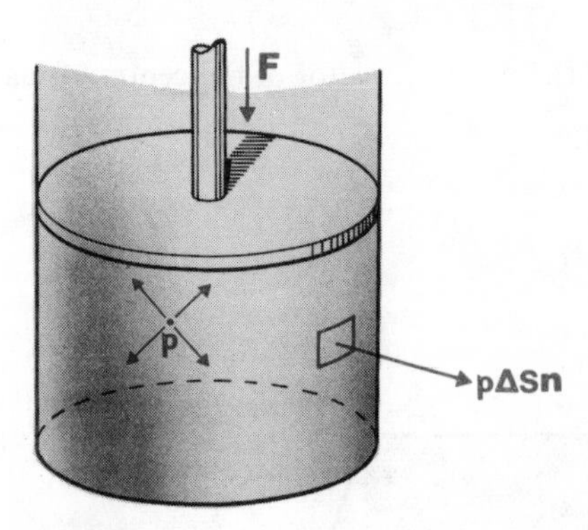

(A cycle pump is an example of this, the air in the pump being the fluid.) The increase in pressure does not cause forces in the same direction as the original force however. Wherever a hole is drilled in the vessel, the fluid will squirt out at the same rate (ignoring gravity for the moment) and normal to the surface at that point. At a boundary of a fluid, the side of the tube for example, the pressure, p, results in a force. The direction of the force is in the direction of the normal to the surface, $\mathbf{n}$; its magnitude is $p\,\Delta S$, where ΔS is the area of the surface being considered.

How is the pressure produced in a fluid? It can be produced by a plunger, as in our example; this is the principle behind hydraulic presses. It can also be produced at points of a fluid in static equilibrium in a gravitational field by the weight of the fluid above. We live at the bottom of an atmosphere of air, and the pressure resulting from the depth of air above us is approximately 10^5 Nm^{-2}. Water is a more dense fluid, and produces substantial pressures in relatively shallow layers. Whatever the fluid, if it is stationary, the pressure at a point is determined by the weight of fluid above that point. If we consider a horizontal surface of cross-sectional area A, at depth h, then the weight of fluid above it would be given by

$$(\text{density}) \times g \times (\text{volume of tube of fluid above surface}).$$

Therefore

$$pA = \rho g h A,$$

i.e.

$$p = \rho g h.$$

The pressure at a point in a fluid in a uniform gravitational field is therefore proportional to the vertical distance of the point beneath the surface of the fluid. Such a pressure is called the *hydrostatic pressure*, and it will give rise to a force acting normally to *any* surface immersed in the fluid at that level, not only the horizontal surface we have been considering.

Now we shall concentrate our attention on those flat surfaces which form the boundaries of the fluid. We shall investigate what the total force on such a surface will be, and where that resultant force can be considered to act.

To answer these questions, let us consider a flat plate of arbitrary shape, with cross-sectional area A, inclined to the horizontal at an angle α. It represents part of the boundary to a liquid of density ρ.

The total force on the element of area ΔA is in a direction perpendicular to the plate and of magnitude $\rho gh\,\Delta A$.

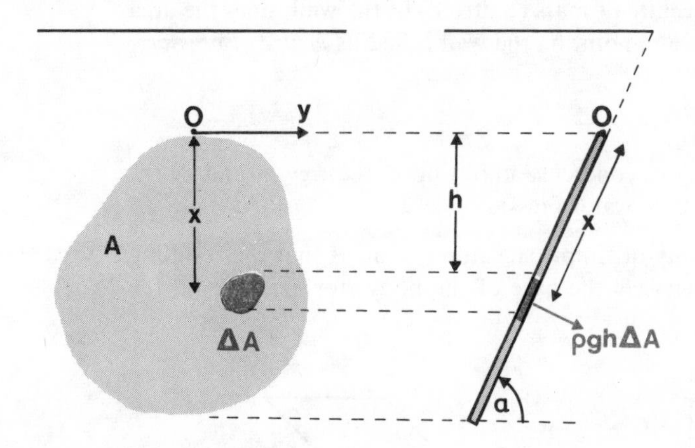

Therefore the total force on the plate is given by

$$\int_A \rho gh\,dA = \int_A \rho gx \sin\alpha\,dA$$

$$= \rho g \sin\alpha \int_A x\,dA$$

$$= \rho g \sin\alpha\,A\bar{x}.$$

In this last step we have used a particular case of the formula for the centre of mass of a flat plate. First, we have taken the x-component only, and second, we have used the fact that the mass of the plate is directly proportional to the area. (Sometimes when the mass of an object is irrelevant to the problem under discussion, as in this case, the centre of mass is called the *centroid* to emphasize its purely geometric property.) Thus the magnitude of the total force on the plate is $\rho g\bar{x}\sin\alpha\,A$, that is,

(hydrostatic pressure at the depth of the centre of mass) × (total area).

Now we must determine the point at which this resultant force can be considered to act. We take moments about some origin O, and assume that the resultant force acts at a point with position vector $\bar{\mathbf{r}}_p$. The unit vector $\mathbf{n}$ is normal to the plate. Then

$$\bar{\mathbf{r}}_p \times (\rho g\bar{x}\sin\alpha\,A\mathbf{n}) = \int_A \mathbf{r} \times \rho gx \sin\alpha\,\mathbf{n}\,dA,$$

so evaluating the vector products and using $\mathbf{i} \times \mathbf{n} = -\mathbf{j}$ etc., we get

$$(-\bar{x}_p\mathbf{j} + \bar{y}_p\mathbf{i})\bar{x}A = \int_A (-x^2\mathbf{j} + xy\mathbf{i})\,dA,$$

where we have substituted $\bar{\mathbf{r}}_p = \bar{x}_p\mathbf{i} + \bar{y}_p\mathbf{j}$. Hence

$$\bar{x}_p = \frac{1}{\bar{x}A}\int_A x^2\,dA, \qquad \bar{y}_p = \frac{1}{\bar{x}A}\int_A xy\,dA.$$

These expressions allow us to determine the position on a boundary surface at which the resultant force due to the pressure of the fluid appears to act. The point is known as the *centre of pressure*.

SAQ 23

Show that for a vertical rectangular plate of length L and width a, which is the side of a vessel filled with liquid, the centre of pressure is $\frac{2}{3}L$ below the top edge, i.e. $\frac{1}{6}L$ below the centre of mass.

(Solution is given on p. 44.)

Summary

Pressures in fluids produce forces on the walls of any container which bounds them. For a fluid at rest with a free surface (neither having a rigid boundary at the top nor being acted upon by atmospheric pressure) the hydrostatic pressure is proportional to the depth below the surface. The magnitude of the resultant force over an area of the wall is equal to the pressure at the centre of mass (centroid) of the wall times the area of the wall. The resultant force acts at a point on the wall called its *centre of pressure.*

6.4.3 The Gravity Dam

These structures are built of masonry or concrete and rely upon the weight and distribution of this material to withstand the water pressure.

The normal condition for the stability of simple structures in air is that the resulting gravitational force on the body intersects the base of the body; for example, in the figure, (a) is stable, (b) is not.

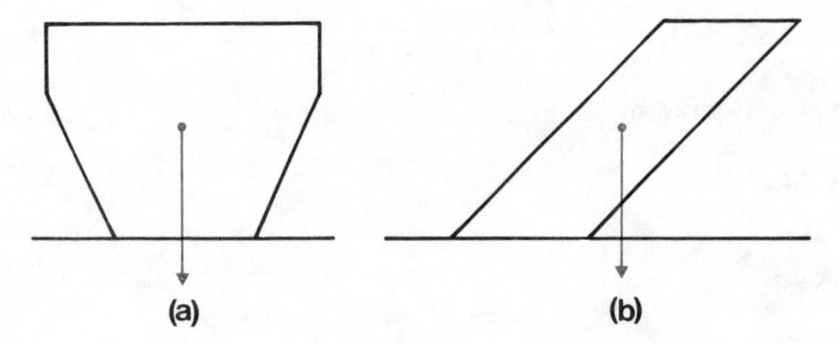

In the case of a gravity dam the design rule is different. First we need to consider the resultant of two forces: the weight acting through the centre of mass, and the force due to the pressure of the water (which is tending to tip the dam over) acting through the centre of pressure. The safety rule that is applied is that this resultant should intersect the middle third of the base of the dam. This is partly in order to counteract additional forces from below.

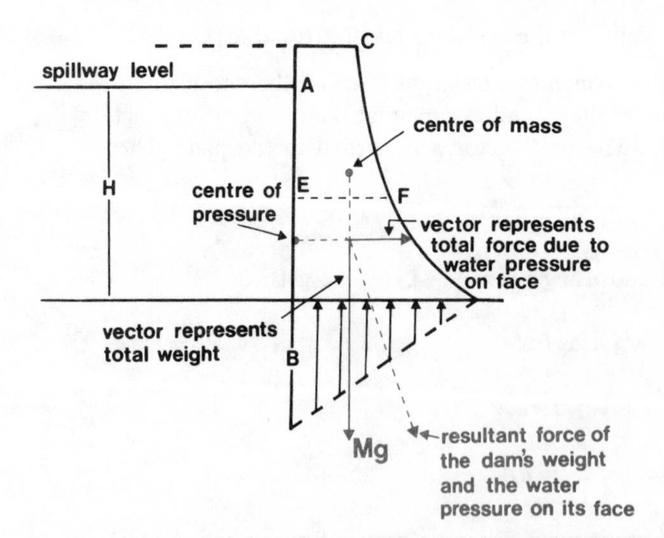

The dam and the surrounding ground are often porous; water penetrates and tends to lift the dam (indicated schematically by the lower arrows in the figure). This effect gives rise to an additional turning moment.

The point E in any horizontal section EF is referred to as the *heel*, and F is called the *toe*. The water pressure at E will depend upon the "head" of water AE; that at F will be zero unless there is a body of water in front of the dam ("tail" water).

For our particular study we shall investigate a dam with a flat front and a trapezoidal cross-section as shown in the figure.

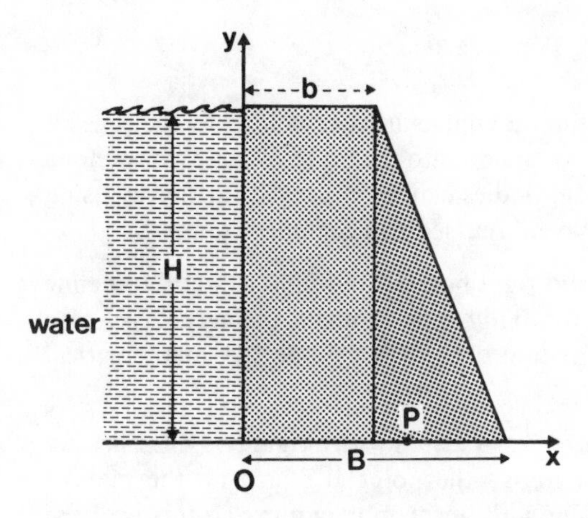

The width of the top, b, is taken to be fixed (usually this is governed by the width of a road along the top). The required height H is also known. The problem is to find the minimum base width B so that the dam conforms to the "middle-third" rule. The forces involved are the resultant force due to the weight of the dam, and the resultant force due to the pressure of the water on the vertical back of the dam. The density of the material of the dam is denoted by ρ_m ; ρ denotes the density of water.

SAQ 24

Translate the "middle-third" rule into a mathematical condition on the point P at which the resultant force intersects the base. What do you then know about the sum of the moments of the two forces about P?

(Solution is given on p. 44.)

SAQ 25

Show that the x-co-ordinate of the centre of mass is

$$\bar{x} = \frac{1}{3}\frac{B^2 + bB + b^2}{B + b}$$

(Solution is given on p. 44.)

SAQ 26

From SAQ 23, the y-co-ordinate of the centre of pressure is $\frac{H}{3}$. What is the total force that is supposed to act at the centre of pressure on the vertical back of the dam for each vertical strip of width 1 metre?

(Solution is given on p. 45.)

SAQ 27

Now you are in a position to determine the minimum base width. Show that B is to be found by solving the quadratic equation

$$B^2 + bB - b^2 - \frac{\rho}{\rho_m} H^2 = 0.$$

(Solution is given on p. 45.)

Putting some realistic numbers into the expression, say $H = 60\text{m}$, $b = 6\text{m}$, $\frac{\rho_m}{\rho} = 2.4$, gives $B \simeq 37\text{m}$, showing that the base would be approximately six times as wide as the top.

Summary

In this final section we have investigated the stability condition for gravity dams and asked you to derive the formula in a particular case. In practice the front of the dam is often curved and the actual calculations are more complicated, but the principle remains the same.

6.5 SUMMARY OF THE TEXT

A proper study of rigid bodies would require a course at least as long as the present one; in this text we have made only a brief incursion into the topic. What we have done is to demonstrate that our modelling of real bodies can be improved further by using the idealization of a rigid body, and we have discussed some simple examples.

In studying the rotation of rigid systems and rigid bodies, we found it useful to define a new physical quantity associated with the body—the moment of inertia. We also introduced surface and volume integrals in order to carry out the mathematical manipulations more elegantly and simply.

We have described several important "centres" associated with a rigid body. Earlier, we had introduced the idea of the centre of mass—the point at which all the mass is considered to be concentrated when dealing with linear momentum. To this we have added the concept of the centre of gravity—the point through which the resultant of the gravitational forces on a body are assumed to act. We showed that this can only be usefully defined when the body is in a uniform gravitational field; under these circumstances it coincides with the centre of mass. We then introduced the centre of pressure—the point through which the resulting force due to fluid pressure on a surface of the body may be assumed to act.

Finally we looked at a simple case study of the design of a gravity dam.

6.6 SOLUTIONS TO SELF-ASSESSMENT QUESTIONS

Solution to SAQ 1

A or C, although the phrase "internal forces" in C is redundant since these cancel out in pairs, by Newton's third law.

Solution to SAQ 2

B

Solution to SAQ 3

(i) **0** (The zero *vector*, *not* the zero scalar.)
(ii) **k**, where **k** is a unit vector coming out of the page, obtained by using the right-hand rule.

Solution to SAQ 4

No. If $f(x)$ is negative anywhere in $[a, b]$, we get either a negative area (impossible, by definition) or the incorrect numerical value of the area.

Solution to SAQ 5

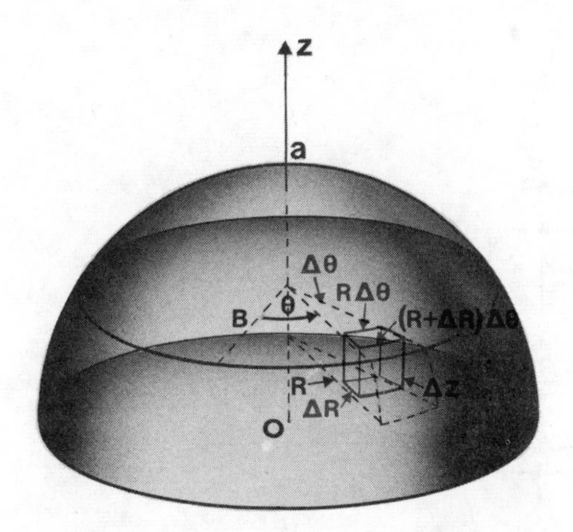

From the figure you can see that the small volume ΔV is shaped rather like the outer part of a slice of cheese with the measurements shown in the figure. The area of the top surface is

$$\tfrac{1}{2}(R + \Delta R)^2 \, \Delta\theta - \tfrac{1}{2}R^2 \, \Delta\theta,$$

because the area of the sector of a circle subtending an angle θ at the centre is $\pi r^2 \dfrac{\theta}{2\pi}$, i.e. $\tfrac{1}{2}r^2\theta$. This is

$$R \, \Delta R \, \Delta\theta + \tfrac{1}{2}(\Delta R)^2 \, \Delta\theta$$

or

$$\Delta R \, \Delta\theta \left(R + \frac{\Delta R}{2}\right).$$

Intuitively, we may ignore $\dfrac{\Delta R}{2}$ compared with R, and so

$$\Delta V = R \, \Delta R \, \Delta\theta \, \Delta z.$$

(If you are unhappy about ignoring the terms, you can put $\Delta R = \dfrac{a}{n}$ and on taking the limits for large n during the R-summation, you will obtain the same result as ours.)

Transcribing this to the integral gives

$$\text{volume} = \int_0^a \left[\int_0^B \left[\int_0^{2\pi} d\theta \right] R\,dR \right] dz.$$

The order in which the integrals occur is important in this case since $B(=(a^2 - z^2)^{1/2})$ is dependent on z and the integration over R must be performed before the integration over z. The steps are as follows:

$$\begin{aligned}
\text{volume} &= \int_0^a \int_0^B 2\pi R\,dR\,dz \\
&= \int_0^a \pi B^2\,dz \\
&= \int_0^a \pi(a^2 - z^2)\,dz = \left[\pi\left(a^2 z - \frac{z^3}{3}\right)\right]_0^a \\
&= \frac{2\pi a^3}{3}.
\end{aligned}$$

Solution to SAQ 6

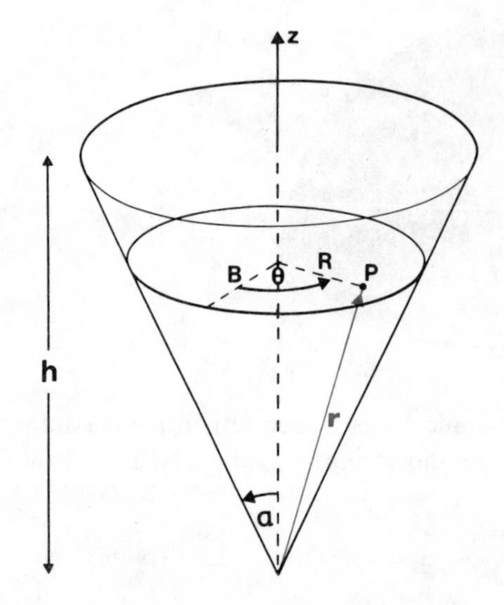

$$\int_V \mathbf{r}\,dV = \int_0^h \left[\int_0^B \left[\int_0^{2\pi} (R\cos\theta\,\mathbf{i} + R\sin\theta\,\mathbf{j} + z\mathbf{k})\,d\theta \right] R\,dR \right] dz$$

$$= \mathbf{k} \int_0^h \int_0^{z\tan\alpha} 2\pi z R\,dR\,dz.$$

(The "symmetry" argument could also be used to show that the **i**- and **j**-components are zero.) Thus

$$\int_V \mathbf{r}\,dV = \pi\mathbf{k} \int_0^h z^3 \tan^2\alpha\,dz = \frac{\pi h^4 \tan^2\alpha}{4}\,\mathbf{k}.$$

Solution to SAQ 7

(i) $$\int_0^4 \int_{y/2}^2 F(x, y)\, dx\, dy$$

(ii) $$\int_{-a}^{a} \int_{-\sqrt{a^2-x^2}}^{+\sqrt{a^2-x^2}} F(x, y)\, dy\, dx$$

and

$$\int_0^a \int_0^{2\pi} G(r, \theta)\, d\theta\, r dr.$$

Solution to SAQ 8

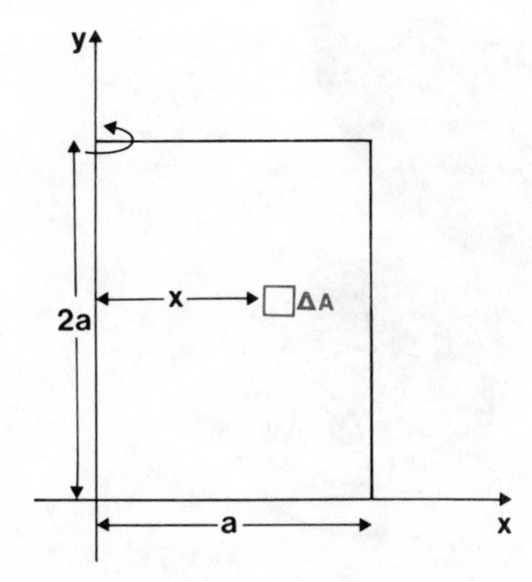

The door is hinged about the y-axis as shown in the above figure.

If ρ is the mass per unit area (i.e. $M = 2a^2\rho$),

$$\begin{aligned}\text{moment of inertia} &= \int_S r^2 \rho\, dA \\ &= \int_0^a \int_0^{2a} x^2 \rho\, dy\, dx \\ &= \int_0^a 2x^2 a\rho\, dx \\ &= 2a\rho \left[\frac{x^3}{3}\right]_0^a \\ &= \frac{2a^4\rho}{3} = \frac{Ma^2}{3}.\end{aligned}$$

Solution to SAQ 9

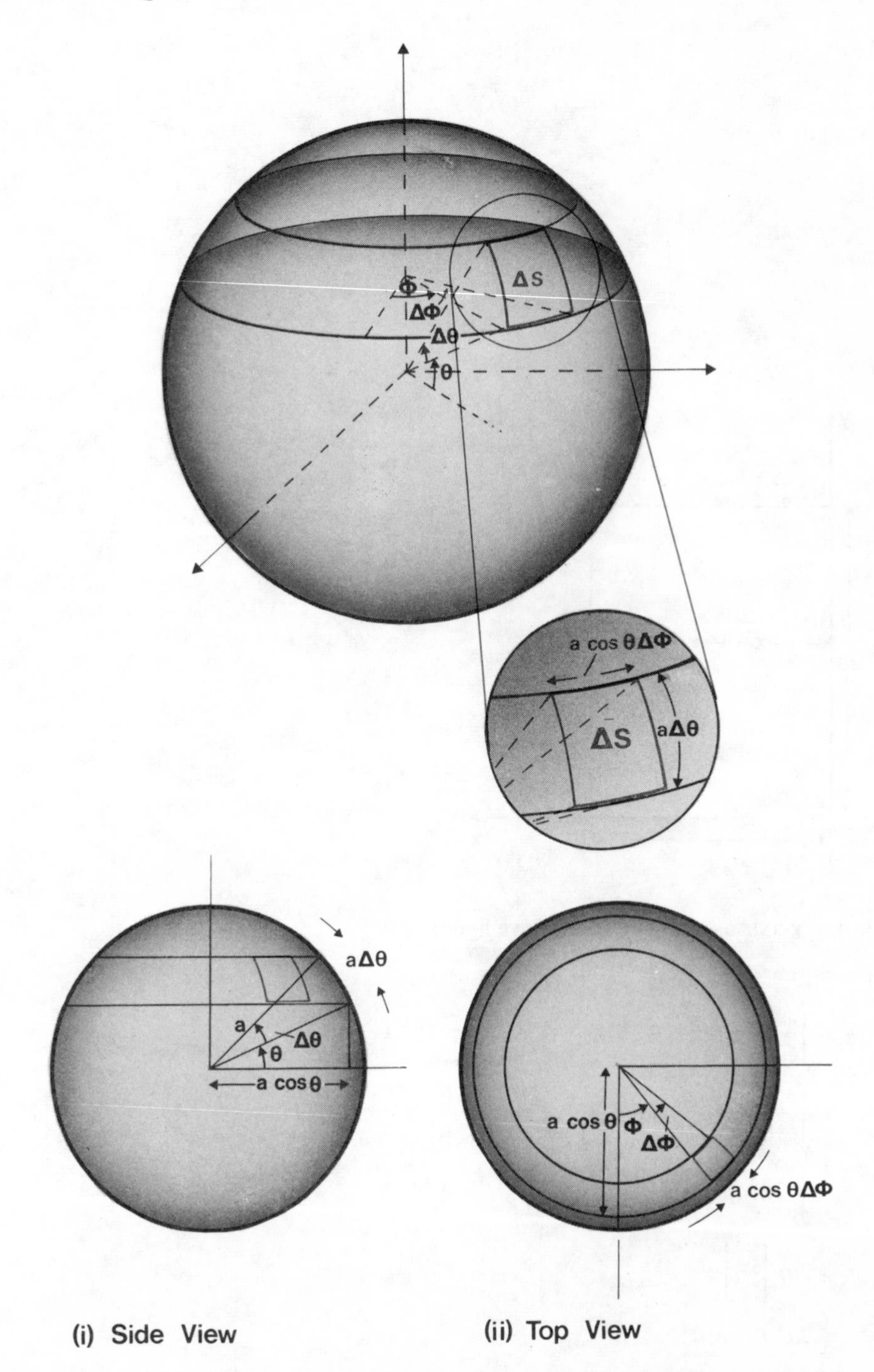

(i) Side View (ii) Top View

Cutting the surface of the sphere into elements as shown in the figure, we obtain

$$\Delta S = a^2 \cos\theta \, \Delta\theta \, \Delta\phi.$$

It follows that

$$\int_S dS = a^2 \int_{-\pi/2}^{+\pi/2} \cos\theta \int_0^{2\pi} d\phi \, d\theta$$

$$= 2\pi a^2 \int_{-\pi/2}^{+\pi/2} \cos\theta \, d\theta = 2\pi a^2 \Big[\sin\theta\Big]_{-\pi/2}^{+\pi/2} = 4\pi a^2.$$

Solution to SAQ 10

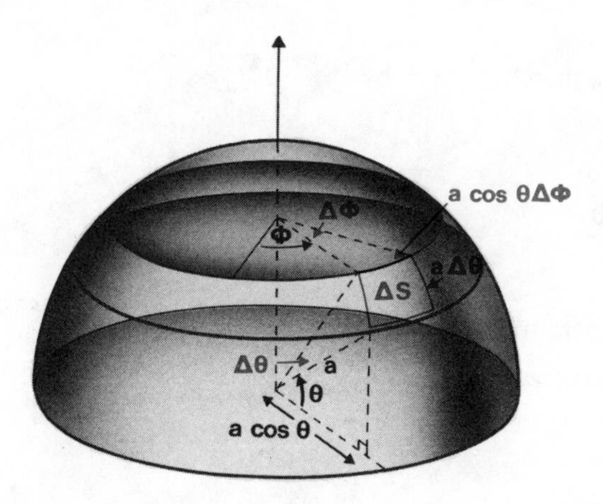

The element of area ΔS is given by

$$\Delta S = a\,\Delta\theta \times a\cos\theta\,\Delta\phi$$

$$\therefore \int_S \mathbf{r}\,dS = \int_0^{\pi/2}\int_0^{2\pi} \mathbf{r}\,a^2\cos\theta\,d\phi\,d\theta$$

$$= \int_0^{\pi/2}\int_0^{2\pi} \mathbf{k}\,a\sin\theta\,a^2\cos\theta\,d\phi\,d\theta$$

$$= \mathbf{k}\int_0^{1}\int_0^{2\pi} a^3\sin\theta\,d\phi\,d(\sin\theta)$$

$$= \mathbf{k}a^3\int_0^1 (2\pi\sin\theta)d(\sin\theta)$$

$$= \pi a^3\mathbf{k}.$$

Solution to SAQ 11

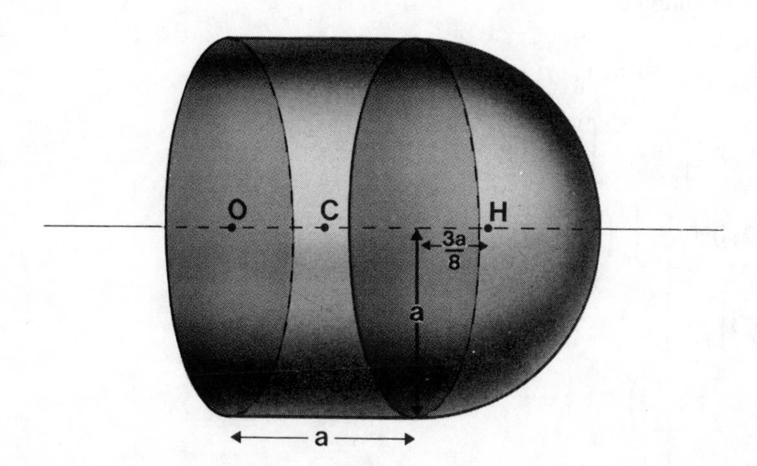

From Example 9, p. S71, the centre of mass of the hemisphere is distance $OH = a + \frac{3a}{8} = \frac{11a}{8}$, along the axis of symmetry from O. The mass of the hemisphere is $\frac{2}{3}\pi\rho a^3$, where ρ is the density.

The centre of mass of the cylinder is distance $OC = \frac{a}{2}$ along the axis of symmetry from O, and the mass is $\pi\rho a^3$.

The centre of mass of the composite body is therefore given by

$$\bar{x} = \frac{\pi\rho a^3 \times \frac{a}{2} + \frac{2}{3}\pi\rho a^3 \times \frac{11a}{8}}{\pi\rho a^3 + \frac{2}{3}\pi\rho a^3} = \frac{17a}{20}$$

from O or distance $\frac{3a}{20}$ from the face of the hemisphere along the axis of the cylinder.

Solution to SAQ 12

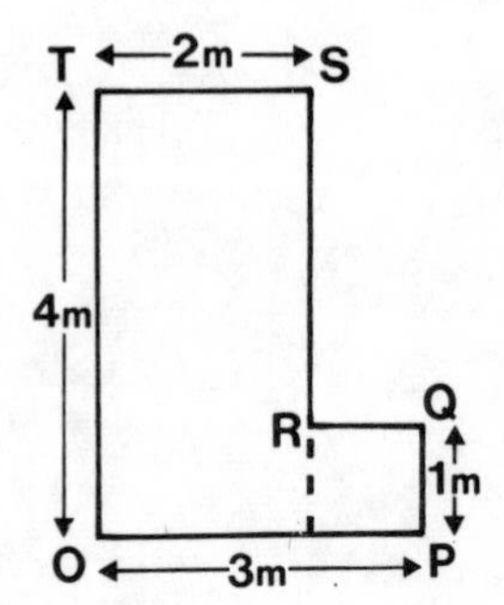

As in SAQ 11, you may assume the positions of the centres of mass of the component parts to determine the position of the overall centre of mass. Here, however, we shall determine it from first principles as an example. Consider a plane section of the lamina, and take ρ as the mass per unit area. Then the position vector of the centre of mass of the lamina, $\bar{\mathbf{r}}$, is defined by

$$M\bar{\mathbf{r}} = \int_A \mathbf{r}\rho \, dA,$$

where A is the total surface area and $M = \rho A$.

Let the point O be the origin, the x-axis be along OP and the y-axis be along OT. Then on dividing the lamina into two sections, as shown by the dotted line, we can set up the required double integral over A as follows

$$\begin{aligned}
\int_A \mathbf{r}\rho \, dA &= \int_0^4 \int_0^2 \rho(x\mathbf{i} + y\mathbf{j}) \, dx \, dy + \int_0^1 \int_2^3 \rho(x\mathbf{i} + y\mathbf{j}) \, dx \, dy, \\
&= \rho\left(\int_0^4 \left[\frac{x^2}{2}\mathbf{i} + yx\mathbf{j}\right]_0^2 dy + \int_0^1 \left[\frac{x^2}{2}\mathbf{i} + yx\mathbf{j}\right]_2^3 dy\right) \\
&= \rho\left(\int_0^4 (2\mathbf{i} + 2y\mathbf{j}) \, dy + \int_0^1 \left(\frac{5}{2}\mathbf{i} + y\mathbf{j}\right) dy\right) \\
&= \rho\left(\left[2y\mathbf{i} + y^2\mathbf{j}\right]_0^4 + \left[\frac{5}{2}y\mathbf{i} + \frac{y^2}{2}\mathbf{j}\right]_0^1\right) \\
&= \rho\left(8\mathbf{i} + 16\mathbf{j} + \frac{5}{2}\mathbf{i} + \frac{1}{2}\mathbf{j}\right) \\
&= \rho\left(\frac{21}{2}\mathbf{i} + \frac{33}{2}\mathbf{j}\right).
\end{aligned}$$

If we equate this to $M\bar{\mathbf{r}}$, where M is the total mass (i.e. 9ρ), we finally obtain $\bar{\mathbf{r}} = \frac{7}{6}\mathbf{i} + \frac{11}{6}\mathbf{j}$.

Solution to SAQ 13

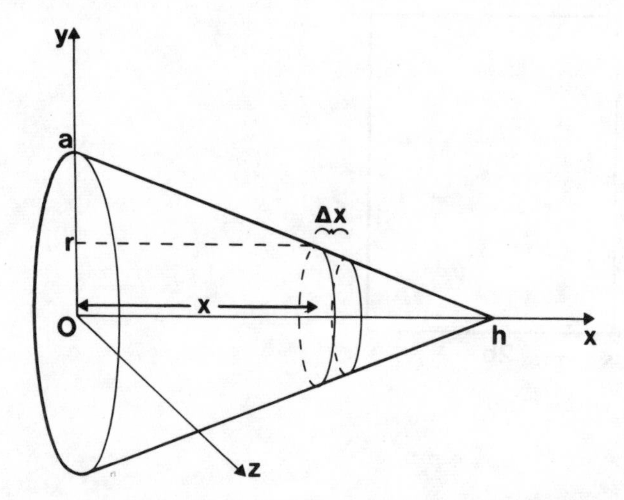

The centre of mass will lie on the axis of symmetry. The mass of the cone is $\rho\pi a^2 h/3$, where ρ is the density. Consider a disc of thickness Δx which is parallel to the circular face of the cone and at distance x from it. The mass of this disc is approximately $\rho\pi r^2 \,\Delta x$, where r is the radius of the disc. By means of similar triangles, we can find r in terms of x; in fact it is quite easy to see that

$$\frac{a}{h} = \frac{r}{h - x},$$

i.e.

$$r = a\left(1 - \frac{x}{h}\right),$$

so the mass of the disc is approximately $\rho\pi a^2\left(1 - \frac{x}{h}\right)^2 \Delta x$. If $\bar{x}$ is the distance of the centre of mass from the face, we have

$$\begin{aligned}
\bar{x}\rho\,\frac{\pi a^2 h}{3} &= \int_0^h \rho\pi a^2\left(1 - \frac{x}{h}\right)^2 x\,dx \\
&= \rho\pi a^2 \int_0^h \left(x - 2\frac{x^2}{h} + \frac{x^3}{h^2}\right) dx \\
&= \rho\pi a^2 \left[\frac{x^2}{2} - \frac{2x^3}{3h} + \frac{x^4}{4h^2}\right]_0^h \\
&= \rho\pi a^2 \left(\frac{h^2}{2} - \frac{2}{3}h^2 + \frac{h^2}{4}\right) \\
&= \frac{\rho\pi a^2 h^2}{12}.
\end{aligned}$$

Hence

$$\bar{x} = \frac{h}{4}.$$

Solution to SAQ 14

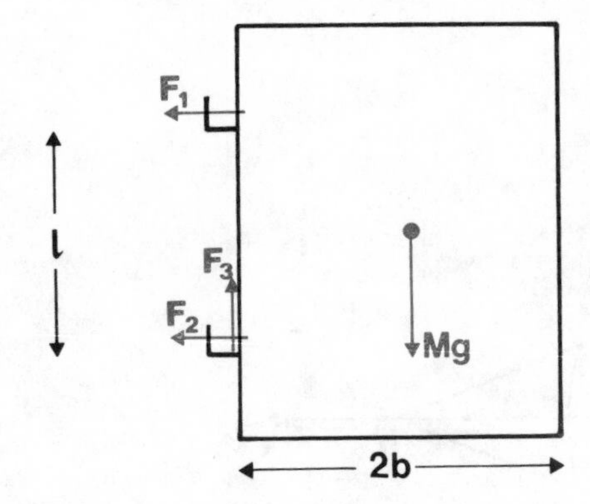

We have

$$F_3 = Mg.$$

Taking moments about the lower hinge, we obtain

$$lF_1 = bMg.$$

Taking moments about the upper hinge, we obtain

$$lF_2 + Mbg = 0.$$

It follows that

$$F_1 = \frac{bMg}{l},\ F_2 = -\frac{bMg}{l}$$

and

$$|\text{resultant at upper hinge}| = \frac{bMg}{l},$$

$$|\text{resultant at lower hinge}| = \frac{Mg}{l}(l^2 + b^2)^{1/2}$$

Solution to SAQ 15

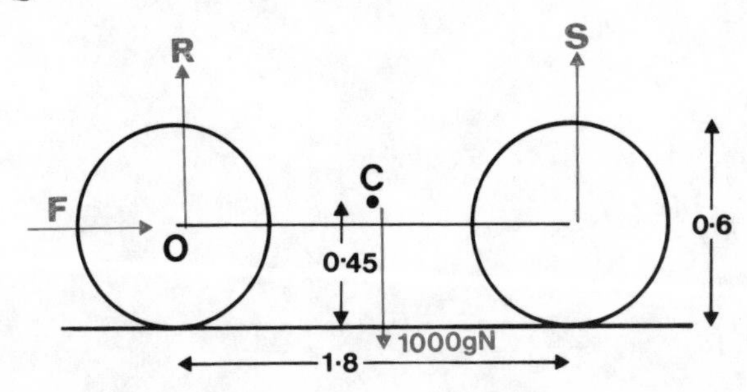

Resolving forces vertically, we obtain

$$R + S = Mg = 1000\,g\text{N}, \qquad \text{(i)}$$

where $g = 9.81\ \text{ms}^{-2}$.

The force F applied at the rear axle = (1000×1) N. Taking moments about the mass centre, we obtain

$$S \times 0.9 + F \times 0.15 = R \times 0.9,$$

hence

$$R - S = \frac{1000}{6}\ \text{N}. \qquad \text{(ii)}$$

From equations (i) and (ii), we have

$$R = \frac{1}{2}\left(1000\,g + \frac{1000}{6}\right)$$

$$= 500\left(g + \frac{1}{6}\right)\ \text{N} \simeq 4990\ \text{N}.$$

From equation (i), it follows that

$$S = 1000\,g - 500(g + \tfrac{1}{6})$$

$$= 500(g - \tfrac{1}{6})\text{ N} \simeq 4820\text{ N}.$$

Solution to SAQ 16

This mathematical model is inappropriate in situations where the bodies are easily deformed on contact and the elastic restoring forces in the body are absent. For example, butter does not bounce—it flattens; a sticky object falling on to the floor may adhere to the floor.

Solution to SAQ 17

Impulse given $= mv = 0.11 \times 10 = 1.1\text{ kg ms}^{-1}$.

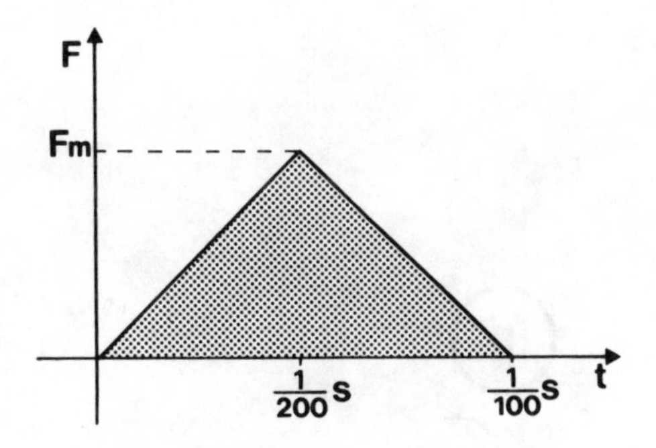

If F_m is the maximum force, then the impulse, $\int F\,dt$, is the area of the triangle on the graph, which is

$$\frac{1}{2} \times \frac{1}{100} \times F_m.$$

Therefore

$$\frac{1}{200} F_m = 1.1,$$

so

$$F_m = 220\text{ N}.$$

Solution to SAQ 18

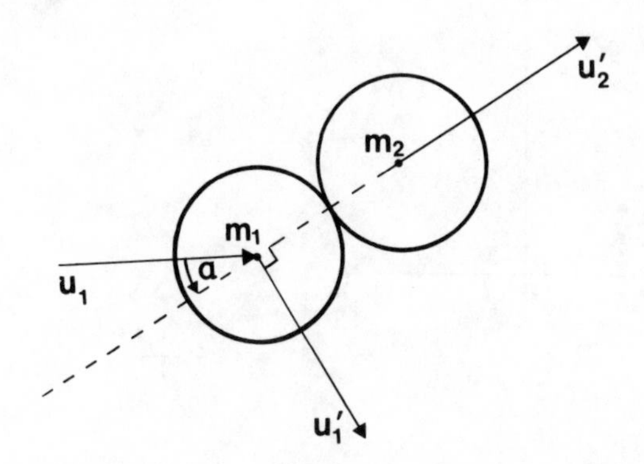

Let the masses of the two spheres be m_1 and m_2, where the sphere that is initially at rest has mass m_2.

The initial speed of m_1 is u_1, and the final speeds of the two spheres are u_1' and u_2'. The stationary sphere receives an impulse along the line joining the centres and so moves off in the direction shown. Because the angle between the two final velocities is

$\frac{\pi}{2}$, u'_1 must be at right-angles to the line joining the centres. With regard to the components of the velocity along this line, Newton's law of restitution gives

$$u'_2 - 0 = -e(0 - u_1 \cos \alpha),$$

which, for $e = 1$, becomes

$$u'_2 = u_1 \cos \alpha. \qquad \text{(i)}$$

Conservation of momentum in the same direction gives

$$m_1 u_1 \cos \alpha + 0 = m_1 u'_1 \cos \frac{\pi}{2} + m_2 u'_2,$$

so

$$m_1 u_1 \cos \alpha = m_2 u'_2. \qquad \text{(ii)}$$

Equations (i) and (ii) can be satisfied simultaneously only if $m_1 = m_2$.

Solution to SAQ 19

a = **0** or **b** = **0**, or **a** is parallel to **b**.

Solution to SAQ 20

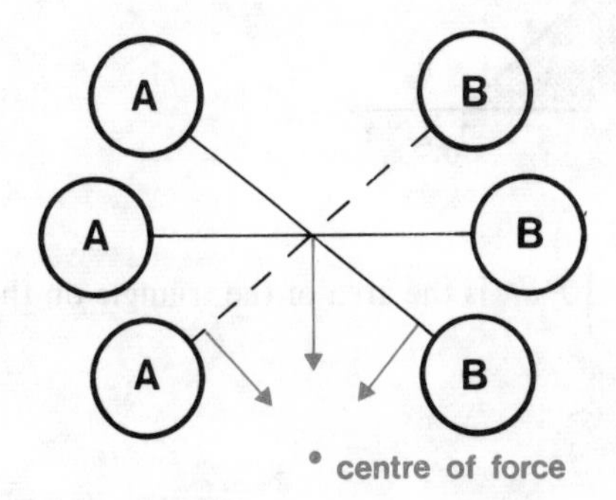

Initially there is a larger gravitational force on B than on A. The resultant force will therefore cross the line joining A and B nearer to B than to A. As the dumb-bell is rotated, the line of action of the resultant force moves through the centre of mass, and then passes nearer to A.

Solution to SAQ 21

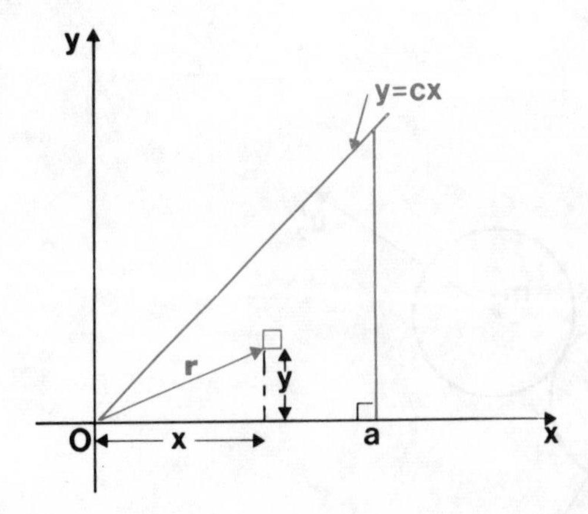

By definition

$$\bar{\mathbf{r}} \int_V \rho \, dV = \int_V \mathbf{r} \rho \, dV.$$

In this case, ρ is constant and the volume integral can be replaced by an integral over the area A. Thus

$$\bar{\mathbf{r}} \int_A \rho \, dA = \int_A \mathbf{r} \rho \, dA,$$

so

$$\bar{\mathbf{r}}\int_0^a\int_0^{cx} dy\, dx = \int_0^a\int_0^{cx}(x\mathbf{i}+y\mathbf{j})\, dy\, dx$$

i.e.

$$\bar{\mathbf{r}}\int_0^a\Big[y\Big]_0^{cx} dx = \int_0^a\left[xy\mathbf{i}+\frac{y^2}{2}\mathbf{j}\right]_0^{cx} dx$$

i.e.

$$\bar{\mathbf{r}}\int_0^a cx\, dx = \int_0^a\left(cx^2\mathbf{i}+\frac{c^2x^2}{2}\mathbf{j}\right) dx.$$

Hence

$$\bar{\mathbf{r}}\frac{ca^2}{2} = \frac{ca^3}{3}\mathbf{i}+\frac{c^2a^3}{6}\mathbf{j}$$

i.e.

$$\bar{\mathbf{r}} = \frac{2a}{3}\mathbf{i}+\frac{ca}{3}\mathbf{j}.$$

Thus, for a right-angled triangle in the position shown, the centre of mass is at a height $\frac{1}{3}$ of the height of the triangle and $\frac{1}{3}$ of the way along the base from the right-angle.

Solution to SAQ 22

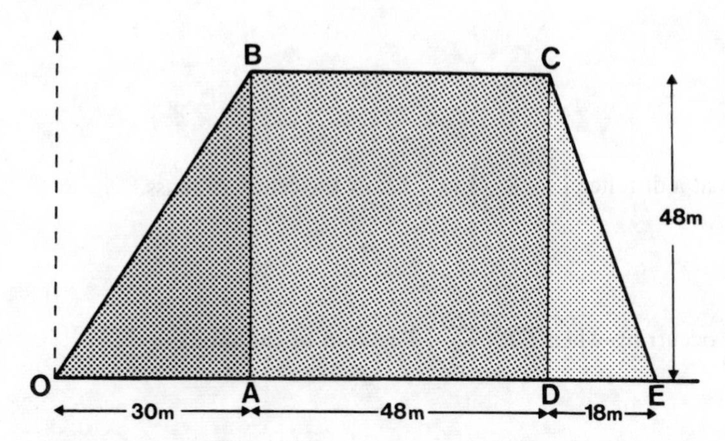

Using the result of SAQ 21, the centre of mass of triangle OAB has position vector $20\mathbf{i}+16\mathbf{j}$. The mass is proportional to $\frac{1}{2}\times 30\times 48$.

The centre of mass of triangle CDE has position vector $84\mathbf{i}+16\mathbf{j}$. Its mass is proportional to $\frac{1}{2}\times 48\times 18$. The centre of mass of square $ABCD$ has position vector $54\mathbf{i}+24\mathbf{j}$. Its mass is proportional to 48×48. Thus the centre of mass of the combined structure is given by

$$\bar{\mathbf{r}} = \frac{\frac{1}{2}\times 30\times 48(20\mathbf{i}+16\mathbf{j})+\frac{1}{2}\times 48\times 18(84\mathbf{i}+16\mathbf{j})+48\times 48(54\mathbf{i}+24\mathbf{j})}{\frac{1}{2}\times 30\times 48+\frac{1}{2}\times 18\times 48+48\times 48}$$

$$\simeq 51\mathbf{i}+21\mathbf{j}.$$

The centre of mass is therefore approximately 21m above the base and 51m along from O.

Solution to SAQ 23

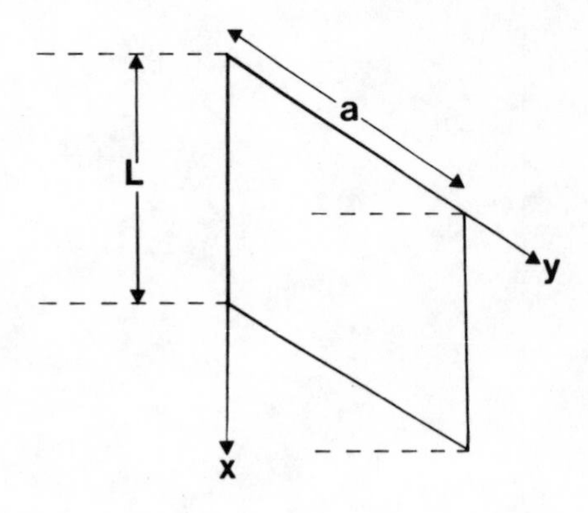

Suppose the width of the plate is a.

The area of the plate is aL, and $\bar{x} = \dfrac{L}{2}$. The equation for $\bar{x}_p$ is

$$\bar{x}_p = \frac{1}{\bar{x}A}\int_A x^2\,dA \qquad \text{(using the result of section 6.4.2)}$$

$$= \frac{1}{L^2a/2}\int_0^L\int_0^a x^2\,dy\,dx$$

$$= \frac{2}{L^2a}\int_0^L ax^2\,dx$$

$$= \frac{2}{L^2}\left[\frac{x^3}{3}\right]_0^L = \frac{2L}{3}.$$

Hence the centre of pressure lies at a distance $\left(\dfrac{2L}{3} - \dfrac{L}{2}\right) = \dfrac{L}{6}$ below the centre of mass.

Solution to SAQ 24

$\dfrac{B}{3} \leqslant OP \leqslant \dfrac{2B}{3}$, the minimum B occurring when $OP = \dfrac{2B}{3}$. The sum of the moments about P is zero.

Solution to SAQ 25

(i) We divide the dam into two sections as shown by the dotted areas in the figure (see p. 31). The centre of mass of the rectangular cross-section has x-co-ordinate $\dfrac{b}{2}$, while that of the triangular cross-section has x-co-ordinate $b + \dfrac{(B-b)}{3}$. The masses of these two sections are proportional to $\rho_m Hb$ and $\rho_m H\dfrac{(B-b)}{2}$ respectively, and hence

$$\bar{x} = \frac{\rho_m H\left[\dfrac{b^2}{2} + \dfrac{B-b}{2}\left(b + \dfrac{1}{3}(B-b)\right)\right]}{\rho_m H\left(b + \dfrac{(B-b)}{2}\right)}$$

$$= \frac{b^2 + Bb - b^2 + \frac{1}{3}(B-b)^2}{2b + B - b}$$

$$= \frac{1}{3}\,\frac{B^2 + bB + b^2}{B + b}.$$

Solution to SAQ 26

Force per metre of the vertical back of the dam

= (pressure at the mass-centre of the plane section) × (area of section)

$$= \rho \frac{gH}{2} \times H \times 1$$

$$= \rho \frac{gH^2}{2}.$$

Solution to SAQ 27

Taking moments about P, the point of intersection of the resultant force and the base, and using the results of SAQ's 24, 25 and 26, we have

$$\underbrace{\left(\frac{2B}{3} - \bar{x}\right)\rho m \frac{H(B+b)g}{2}}_{\text{magnitude of moment of weight}} - \underbrace{\frac{H}{3}\rho g \frac{H^2}{2}}_{\text{magnitude of moment due to water pressure}} = 0.$$

APPENDIX 1

Theorem

The moment, $\mathbf{M}$, and the moment of momentum, $\mathbf{h}$, taken about the centre of mass of a rigid body are related by the equation

$$\mathbf{M} = \frac{d\mathbf{h}}{dt},$$

regardless of how the centre of mass is moving and (possibly) accelerating. (That is, the same form of moment of momentum equation applies both to fixed points and to the centre of mass.)

Proof

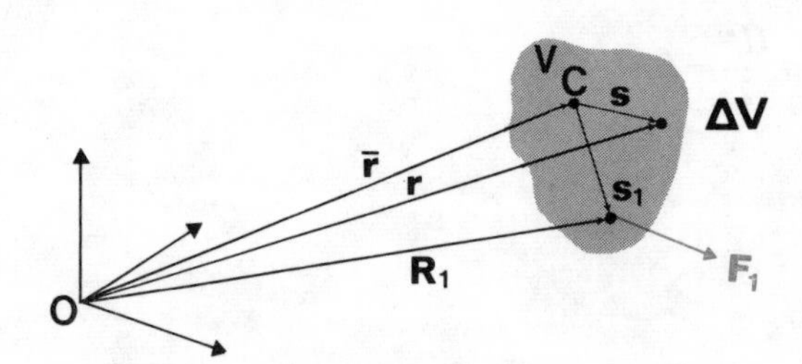

O is a fixed origin. For simplicity, we shall assume that just one external force, $\mathbf{F}_1$, is acting (if there were more, we would carry a $\sum$ through the following equations).

Then the moment of momentum of the rigid body about O, $\mathbf{h}_O$, is by definition,

$$\mathbf{h}_O = \int_V \mathbf{r} \times \rho\dot{\mathbf{r}}\, dV.$$

The moment of momentum about the centre of mass C, $\mathbf{h}_c$, is

$$\mathbf{h}_c = \int_V \mathbf{s} \times \rho\dot{\mathbf{s}}\, dV.$$

By definition of the centre of mass,

$$m\bar{\mathbf{r}} = \int_V \mathbf{r}\, \rho\, dV.$$

Thus

$$\mathbf{0} = \int_V \mathbf{s}\rho\, dV,$$

and from the linear momentum equation we have

$$\mathbf{F}_1 = m\ddot{\bar{\mathbf{r}}}.$$

We now have the basic information.

Because $\mathbf{r} = \bar{\mathbf{r}} + \mathbf{s}$, we have, from the definition of $\mathbf{h}_O$,

$$\mathbf{h}_O = \int_V (\bar{\mathbf{r}} + \mathbf{s}) \times (\dot{\bar{\mathbf{r}}} + \dot{\mathbf{s}})\rho\, dV$$

$$= \int_V \bar{\mathbf{r}} \times \dot{\bar{\mathbf{r}}}\rho\, dV + \int_V \mathbf{s} \times \dot{\bar{\mathbf{r}}}\rho\, dV + \int_V \bar{\mathbf{r}} \times \dot{\mathbf{s}}\rho\, dV + \int_V \mathbf{s} \times \dot{\mathbf{s}}\rho\, dV.$$

The last term is $\mathbf{h}_c$. We now manipulate the other terms noting that both $\bar{\mathbf{r}}$ and $\dot{\bar{\mathbf{r}}}$ are constants for the volume integration. We obtain

$$\mathbf{h}_O = m\bar{\mathbf{r}} \times \dot{\bar{\mathbf{r}}} + \left(\int_V \mathbf{s}\rho\, dV\right) \times \dot{\bar{\mathbf{r}}} + \bar{\mathbf{r}} \times \left(\int_V \dot{\mathbf{s}}\rho\, dV\right) + \mathbf{h}_c.$$

Now

$$\int_V \mathbf{s}\rho\, dV = \mathbf{0}$$

and

$$\frac{d}{dt}\left(\int_V \mathbf{s}\rho \, dV\right) = \mathbf{0}.$$

Intuitively, since ρ and the volume integral are both independent of t, we can differentiate inside the integral sign to obtain

$$\int_V \dot{\mathbf{s}}\rho \, dV = \mathbf{0}.$$

(To prove this, we would need to revert to the definition of the volume integral as the limit of a sum.)

Thus

$$\mathbf{h}_O = m\bar{\mathbf{r}} \times \dot{\bar{\mathbf{r}}} + \mathbf{h}_c$$

and

$$\frac{d\mathbf{h}_O}{dt} = m\dot{\bar{\mathbf{r}}} \times \dot{\bar{\mathbf{r}}} + m\bar{\mathbf{r}} \times \ddot{\bar{\mathbf{r}}} + \frac{d\mathbf{h}_c}{dt}.$$

Since for any vector $\mathbf{a}$, $\mathbf{a} \times \mathbf{a} = \mathbf{0}$,

$$\frac{d\mathbf{h}_O}{dt} = \bar{\mathbf{r}} \times m\ddot{\bar{\mathbf{r}}} + \frac{d\mathbf{h}_c}{dt}$$

$$= \bar{\mathbf{r}} \times \mathbf{F}_1 + \frac{d\mathbf{h}_c}{dt}.$$

But

$$\frac{d\mathbf{h}_O}{dt} = \mathbf{R}_1 \times \mathbf{F}_1$$

$$= (\bar{\mathbf{r}} + \mathbf{s}_1) \times \mathbf{F}_1$$

$$= \bar{\mathbf{r}} \times \mathbf{F}_1 + \mathbf{s}_1 \times \mathbf{F}_1.$$

Because

$$\mathbf{s}_1 \times \mathbf{F}_1 = \frac{d\mathbf{h}_c}{dt},$$

we have, finally,

$$\frac{d\mathbf{h}_O}{dt} = \bar{\mathbf{r}} \times \mathbf{F}_1 + \frac{d\mathbf{h}_c}{dt}.$$

Thus the theorem is proved, since we note that nowhere was any restriction put on the actual motion of the centre of mass.

APPENDIX 2

Additional Notes for Smith and Smith, *Mechanics*

In this appendix we are providing some additional explanatory notes for the parts of the set book we have asked you to read.

If you have found any parts difficult, you should turn to the appropriate page reference of S. These page and line references are on the left-hand side of the pages of this appendix. In some cases we indicate, for example, where proofs can be found in other books, and in other cases we add some further explanation of our own.

Page 69, Example 8

In this example, note that r is not the distance from the origin to the shell (i.e. $|\mathbf{r}|$), but is used to represent the horizontal displacement of the shell.

Page 69, line 13

$$\sec^2 \alpha = 1 + \tan^2 \alpha.$$

Page 69, line −14

In the quadratic equation

$$Ax^2 + Bx + C = 0,$$

the discriminant is defined as

$$B^2 - 4AC.$$

The equation has two distinct real roots if the discriminant is positive.

Page 70, last line

This definition of the mass centre is a direct extension of the definition given on page 56, line 25. We assume that the rigid body is made up of a large number of small volumes of mass Δm_i so that we can define

$$\bar{\mathbf{r}} = \frac{\sum \mathbf{r}_i \, \Delta m_i}{\sum m_i}.$$

But $\Delta m_i = \rho_i \, \Delta V_i$, assuming constant density, ρ_i, throughout the elementary volume ΔV_i. Hence

$$\bar{\mathbf{r}} = \frac{\sum \mathbf{r}_i \rho_i \, \Delta V_i}{\sum \Delta m_i}.$$

If M is the total mass, then

$$M = \sum \Delta m_i,$$

and therefore

$$M\bar{\mathbf{r}} = \lim \sum \mathbf{r}_i \rho_i V_i$$
$$= \int_V \mathbf{r}\rho \, dV.$$

Page 71, Example 9

By using symmetry, the volume integral, which is equivalent to a triple integral, can be reduced to a single definite integral.

Page 72, line 8

This definition of linear momentum is an extension of the equation for a system of particles, given on page 56, line −2.

Page 72, Equation (18)

It has also been assumed that ρ is not dependent on time—a reasonable assumption to make for a rigid body.

Page 72, Example 10

When it is said that the wheel rolls, it implies that it rolls *but does not slip*, so that in this case we can deduce that the angular speed $\omega = \dfrac{U}{a}$.

Page 73, line 3

The integral $\int_V \dot{\mathbf{r}}\rho \, dV$ is evaluated as a *double*, not a triple, integral: we integrate first with respect to θ, then with respect to R; this is because we may assume that the wheel has negligible thickness, and ρ is in units of mass per unit area.

Page 75, line −7

$\mathbf{h} = \int_V \mathbf{r} \times \dot{\mathbf{r}}\rho \, dV$ is a direct extension of the equation $\mathbf{h}_O = \sum_{i=1}^{n} m_i \mathbf{r}_i \times \dot{\mathbf{r}}_i$ on page 58, line −4.

Page 76, line 1

Each of the vector equations gives 3 scalar equations when we equate components.

Page 76, line 19

The definition of the mass-centre is given on page 70, last line.

Page 76, line −8

$$\begin{aligned}\frac{d\mathbf{h}}{dt} &= \frac{d}{dt}(M\bar{\mathbf{r}} \times \dot{\bar{\mathbf{r}}}) \\ &= M\dot{\bar{\mathbf{r}}} \times \dot{\bar{\mathbf{r}}} + M\bar{\mathbf{r}} \times \ddot{\bar{\mathbf{r}}}.\end{aligned}$$

The first term on the right-hand side of this equation is zero, so

$$\frac{d\mathbf{h}}{dt} = M\bar{\mathbf{r}} \times \ddot{\bar{\mathbf{r}}}.$$

It follows from Equation (19), page 74, that

$$\frac{d\mathbf{h}}{dt} = \bar{\mathbf{r}} \times \mathbf{F}.$$

Page 76, line −6

This uses the equation on page 74, line 3.

Page 76, last paragraph

This follows from the fact that $\mathbf{r} - \bar{\mathbf{r}}$ is a position vector relative to the mass-centre, and therefore the left-hand side of the last equation represents the moment of the applied forces *about the mass-centre*.

Page 81, line 8

Newton's law of restitution is an empirical law and, unlike the three basic laws of dynamics, it is only an approximation to the true behaviour during a collision.

Page 84, line 3

"Mean tension" means the average value of the tension, i.e.

$$\bar{T} = \tfrac{1}{2}\int_0^2 T \, dt.$$

Glossary

Terms defined in this glossary are printed in CAPITALS. Some terms not defined in this list may be found in the *Mathematical Handbook*.

Term	Definition	Page
ANGULAR MOMENTUM (OR MOMENT OF MOMENTUM) ABOUT A POINT	The ANGULAR MOMENTUM of a rigid body ABOUT A POINT O is defined by $\mathbf{h}_O = \int_V \mathbf{r} \times \dot{\mathbf{r}} \rho \, dV$ where $\mathbf{r}$ is the position vector (referred to O) of a typical point of the body, ρ is the density of the body, and the integration is performed throughout V, the volume of the body.	7
BODY FORCES	BODY FORCES are forces which apply to every point of a body, as for example the force of gravity. They are defined in this text for rigid bodies, but also occur in bodies of deformable materials (p. S73).	
APPLIED SURFACE FORCES	APPLIED SURFACE FORCES are forces which act only on the surface of bodies (both rigid and deformable). (p. S73)	
CENTRE OF GRAVITY	The CENTRE OF GRAVITY of a body is the point through which the resultant gravitational force on the body may be supposed to act.	24
CENTRE OF MASS	See MASS-CENTRE.	7
CENTRE OF PRESSURE	The CENTRE OF PRESSURE of a body is the point on its surface at which the force on the surface due to fluid PRESSURE may be supposed to act.	28
CENTROID	The CENTROID of a body is the same as its MASS-CENTRE. The term is used when the mass of the body is irrelevant to the discussion.	29
COEFFICIENT OF RESTITUTION	See NEWTON'S LAW OF RESTITUTION.	22
ELASTIC COLLISIONS	ELASTIC COLLISIONS are those for which the COEFFICIENT OF RESTITUTION is 1.	23
HYDROSTATIC PRESSURE	HYDROSTATIC PRESSURE is the PRESSURE produced at a point in a fluid by the weight of the fluid above that point.	28
IMPULSE OR IMPULSIVE FORCE	The IMPULSE on a rigid body is the vector difference of the initial and final LINEAR MOMENTUM arising from a collision.	23
IMPULSIVE MOTION	IMPULSIVE MOTION arises as a result of a collision during which IMPULSIVE FORCES are supposed to act.	22
LINEAR MOMENTUM	If $\mathbf{r}$ is the position vector of a typical point in a rigid body of density ρ, the LINEAR MOMENTUM is defined by the VOLUME INTEGRAL $\int_V \dot{\mathbf{r}} \rho \, dV$, the integration being performed throughout the volume V of the body.	7

		Page
MASS-CENTRE	The MASS-CENTRE of a rigid body is the point whose position vector $\bar{\mathbf{r}}$ is defined by $$M\bar{\mathbf{r}} = \int_V \mathbf{r}\rho\, dV.$$ It is the point at which we may suppose all the external forces to act when we consider the LINEAR MOMENTUM of a rigid body.	7
MIDDLE THIRD RULE	The MIDDLE THIRD RULE is a design rule used in the construction of dams. It states that the resultant of the weight (acting through the dam's CENTRE OF MASS) and the force due to water PRESSURE (acting at the dam's CENTRE OF PRESSURE) must pass through the MIDDLE THIRD of the dam's base.	30
PRESSURE	PRESSURE is the scalar force per unit area.	28
NEWTON'S LAW OF RESTITUTION	When two bodies collide, their relative parting velocity in the direction of the common normal is $(-e)$ times their relative approach velocity in the same direction. e is a positive constant called the COEFFICIENT OF RESTITUTION.	23
SURFACE INTEGRALS	If f is a function having the set of points in 3-dimensional space as its domain, and S is a surface, the SURFACE INTEGRAL $$\int_S f(p)\, dS \quad \text{is} \quad \lim_{\|\Delta\| \to 0} \sum f(p_i)\, \Delta S_i$$ provided the limit exists, where $\|\Delta\|$ is the maximum width of ΔS_i.	16
VOLUME INTEGRAL	If g is a function having the set of points of 3-dimensional space as its domain and V is a volume in the space, then the VOLUME INTEGRAL $$\int_V g(p)\, dV \quad \text{is} \quad \lim_{\|\Delta\| \to 0} \sum g(p_i)\, \Delta V_i,$$ provided the limit exists, where $\|\Delta\|$ is the maximum width of ΔV_i.	11

MECHANICS AND APPLIED CALCULUS